Frontiers of Physics

FRONTIERS
OF
PHYSICS

Samuel Gueller

VANTAGE PRESS
New York / Washington / Atlanta
Los Angeles / Chicago

Published by Vantage Press, Inc.
516 West 34th Street, New York, New York 10001

Manufactured in the United States of America
ISBN: 0-533-06757-X

Library of Congress Catalog Card No.: 85-90261

To my wife, Bertha F.,
and sons, Daniel H. and Eduard J.

CONTENTS

Frontiers of Physics

Part One

The Unified Field

INTRODUCTION

On the assumption that by the age of fifty, one has arrived at a certain level of maturity, I decided to write this essay, having recently arrived at the half-century.

That certain maturity is undoubtedly required and my own may well not be sufficient to the cause; the matters I wish to treat are among the more difficult ones for we humans to grasp. This is, in large measure, due to a lack of understanding of their nature, which, in turn, makes them so attractive and challenging. One might say they keep our life in ongoing suspense, so to speak. The intellectual history of all civilized peoples turns about just the themes I wish to touch upon.

In this vein, I want to present a physico-philosophic view in trying to give the clearest possible description of the subjects I shall broach. First, I shall approach them on a case-by-case basis and subsequently as a whole, all in search for homogeneity and to eliminate possible contradictions that may appear to arise.

In my reading over the years, I have noticed that virtually everything known has been written more than once. As a result, the contributions of each author tend to be relatively small and consist of a certain element of originality, in general, representing but a small percentage of the amount actually written. In this essay, I have been no exception to this trend.

While I don't have a great deal to say, it has been necessary, in this essay format, to present the only part that is mine—namely, the new concepts—in a cultural-historical setting. This is our natural way of developing culturally. The parameters in question are general concepts, philosophical principles, accepted theories, and particular concepts, such as God, the present structure of the sciences, and suchlike.

Since all these aspects are well-known facts, I have avoided quoting authors and text locations so as not to sidetrack the reader's attention from the central theme. This is what I hope will capture the reader's interest and attention, being, as it is, the significant part of the essay and expressed, on occasion, with some economy of words.

From the beginning of civilization, men have been trying to explain reality and its phenomena. So it was that the Greeks, when their thought began to develop and they formalized their philosophy and metaphysics, had to define reality and its phenomena. Since then, philosophy has adapted to the new knowledge emanating from the experimental sciences—mainly mathematics and physics—thus giving rise to updated versions of the original, all in a continual effort to explain the world around us. The updated versions, always signifying some progress, run parallel to, or close on the heels of, ongoing discovery.

In view of the above, it is perhaps opportune after the advances of the past century and a half to try to graft the meaning of new discoveries onto the present body of knowledge with a view to improving it, if possible.

This is not a didactic approach to the frontiers of physics—or just how far they may have been pushed—but rather an effort to glimpse beyond, to see the directions in which the next steps may lead.

Cincinnati, November 1985

Chapter I

DEFINITIONS; ESTIMATION OF DOMAIN DIMENSIONS

We shall define the domain of the whole universe as U. By the whole universe we mean the entirety of the region within which all the laws we employ are valid; such laws would be those of physics, biology, and suchlike, along with Newton's law of universal attraction and the geometries, both Euclidean and non-Euclidean—in sum, all those laws that do not predominate at the atomic, molecular, or ionic level. In this domain, which we shall define as e, the laws of electromagnetism, thermodynamics, quantum mechanics, and suchlike apply.

The subatomic domain, where, because of our ignorance of the applicable laws, we have no such laws to apply, we shall define as ee.

If we move to the other extreme—namely, beyond the limits of the universe U as it is known to cosmogeny—then we need a further designation, and we shall employ UU.

Whilst for the domains UU and ee, we have no principles or laws of phenomena insofar as we are unaware of what may take place there, I want to take some inadequately explained phenomena in the domain U and e with a view to inferring what may be going on in ee. In the same fashion and as a function of what is happening in ee, e, and U, we should be able to begin to have some idea of what happens in UU.

In Figure 1 we have a diagramatic representation of the domains in which the visible world is the framework U; the hydrogen atom brings

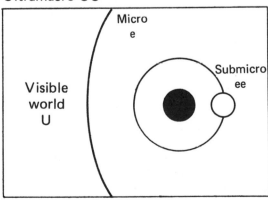

Figure 1

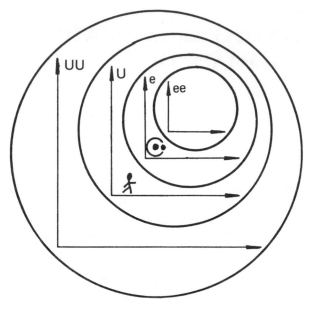

Figure 2

us to the domain e, and the electron to the domain ee. We thus situate the ultramacro domain UU beyond the limits of the universe.

There is a further way of representing these domains, represented in Figure 2, in which the sequence ee, e, U, UU is shown with each domain contained in the next following one, and the axes represent the frame of reference of the phenomena corresponding to each domain.

We shall see how disconcerting it can be to look at phenomena that, while happening in one domain, and overflowing, as it were, are interpreted from the standpoint of a system beyond their domain.

ESTIMATION OF DIMENSIONS BETWEEN U AND e

We are all familiar with the idea of Einstein in his youth, wanting to see how the universe would appear to one sitting on a ray of light.

We now know how this gave rise to a deeper study of time, the resultant modification of almost all the laws of physics, the explanation of some phenomena such as the shifting of Mercury's perihelion, and the discovery of others.

What we propose is a similar "trip," during which there will be a notable relativization of space. This trip will be one of diminishing ourselves

6

progressively until we have the same relationship in size to an electron as we normally do to our planet, and we shall endeavor to glimpse what goes on down there in the microcosmos. We will also see a rapid reduction in linear dimensions.

In order to study a dimensional relationship between the domains U and e, we shall employ estimates of lengths and masses in line with classical physics. If we try the same process with time, however, we must stop and ask ourselves whether relativistic formulae should be applied.

In principle, I think the answer to this is no, since (a) the contraction of lengths demands an extremely high-speed displacement, which, here, is not the case, and (b) as an object is reduced in size, the rate of reduction with respect to time in U can be measured using an accurate chronometer. There comes a moment, of course, when it crosses the frontier from U into e, where there will be a local reference time of e.

We want to find a relationship between the local time T of U with respect to t of e.

For this purpose, we shall assume that the domains are related through a reduction in the time variable in the e level with respect to values found in U; hence, phenomena in e observed from U take place faster, and vice versa.

To this end, we shall assume a linear relationship in the reduction similar to that of the period of a pendulum when its length is reduced. Thus, if we take a pendulum of length L, where L = 30 cm with a period of 1 second, then we will see that the period will be but a tiny fraction of a second if we progressively reduce the length to a value, for example, of L, where L = 0.5 x 10^{-8} cm.

The expressions would be, in U, T = 2 π $\sqrt{(L/g)}$, and in e, t = 2 π $\sqrt{(l/g)}$; then

$$\frac{T}{t} = \frac{L}{l}$$.

Now, if we take L = 298 x 10^{11} cm (diameter of the orbit of the earth around the sun) and l = 0.53 x 10^{-8} cm (diameter of the orbit of the electron around the atomic nucleus), we then have

$$\frac{T}{t} = \frac{300 \times 10^{11}}{\frac{1}{2} \times 10^{-8}} = 300 \times 2 \times 10^{11} \times 10^{8} = 6 \times 10^{21}.$$

According to this approach, the phenomena in e would be some 6 x 10^{21} times faster than in U, measured from U.

Let us consider how much faster time is in e than in U. If we take the period of the earth's revolution about the sun ($T = 3600 \times 24 \times 365 = 31.5 \times 10^6$ seconds), we have

$$t = \frac{T}{6 \times 10^{21}} = \frac{31.5 \times 10^6 \text{ sec}}{6 \times 10^{21}} = 5 \times 10^{-15} \text{ seconds.}$$

Since we know the diameter of the hydrogen electron's orbit and hence the circumference, we can compute the corresponding electron revolutions per second as observed from U and arrive at a value for the speed of the electron

$$l = \pi D = 3.14 \times 0.5 \times 10^{-8} \text{ cm} = 1.57 \times 10^{-8} \text{ cm.}$$
$$v_e = \frac{l}{t} = \frac{1.57 \times 10^{-8} \text{ cm}}{5 \times 10^{-15} \times 1 \text{ sec.}} = 0.30 \times 10^{-8} \times 10^{15}$$
$$= 0.30 \times 10^7 \text{ cm/sec} = 30 \text{ km/sec.}$$

The speed of light is 300,000 km/sec, and the ratio of the above value is 300,000/30—that is, 10,000. If the photon constitutes an "electron" from the domain ee, is this value also the speed relationship between the domains e and ee?

Considering now the speed relationship between U and e, we have the speed of the earth v_T where

$$v_T = \frac{l}{t} = \frac{\pi D}{3 \times 10^7 \text{ sec}} = \frac{3.14 \times 3 \times 10^8}{3 \times 10^7 .} = \frac{9.42 \times 10^8}{3 \times 10^7}$$
$$= \frac{0.94 \times 10^9 \text{ Km}}{3 \times 10^7 \text{ sec}} = \frac{0.94 \times 10^{12} \text{ m}}{3 \times 10^7 \text{ sec}} = 0.30 \times 10^7 \frac{\text{cm}}{\text{sec}},$$

and the speed of the electron, v_e where $v_e = 0.30 \times 10^7$ cm/sec. Thus we have the relation

$$\boxed{\frac{v_e}{v_T} = \frac{0.3 \times 10^7}{0.3 \times 10^7} \cong 1}$$

To a close approximation, the two speeds are equal. Considering the values taken and the consequences of such a result, this is something remarkable.

The U/e domain passage constant for time would be 6×10^{21} and the velocity (domain passage) constant would be unity.

This value shows that if mass, and hence energy, is constant between domains, then velocity must also be constant since it is a factor of momentum.

We will see in Chapter III that the velocity (domain passage) constant between e and ee is also unity.

I will go further and state my belief that there is an absolute time encapsulating local domain times where relativity applies—and these times are related to each other by a constant.

Now, looking at mass, that of the earth (M_T) is 5.98 x 10^{27} grams, while that of the electron (m_e) is 9.11 x 10^{-28} grams; hence the ratio would be

$$\frac{M_T}{m_e} = \frac{5.98 \times 10^{27}}{9.11 \times 10^{-28}} = 0.54 \times 10^{57}.$$

Looking at distance in the same way,

$$\frac{\text{Diameter of earth's orbit}}{\text{Diameter of electron orbit}} = \frac{2 \times 1.5 \times 10^{13} \text{ cm}}{5.3 \times 10^{-9} \text{ cm}} = 0.56 \times 10^{22}.$$

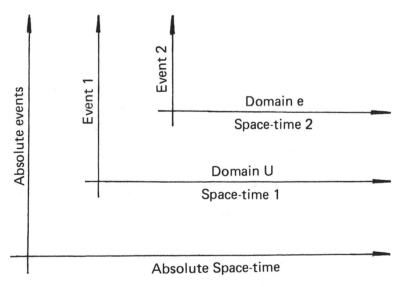

Figure 3

9

The relativistic space-time relationships are applied to phenomena in a single domain, but the laws applicable to phenomena in one domain, when those phenomena pass to another, appear different as a result of the change of scale. In fact, it is as though we were actually using relativistic space. This can be represented diagramatically as in Figure 3.

Events 1 and 2 could, with appropriate corrections, be shifted to an absolute system; and if, as illustrated, they belong to different domains, then, with appropriate corrections, a relationship can be found between them.

ESTIMATION OF DIMENSIONS: DOMAINS e AND ee

We saw that the velocity ratio of the electron to the photon was 1; we shall now look at the corresponding mass, energy, and distance relationships.

The Mass Ratio

The mass, m_e, of the electron would be 9.11×10^{-31} kilograms, or 9.11×10^{-28} grams.

For a value of the mass of the photon, we may take the general expression $\lambda = h/mv$, whence $m_f = h/v\lambda$, where m = mass, m_f = photon mass, v = velocity, λ = wavelength, and h (Planck's constant) = 6.62×10^{-34} joule/second.

For the λ value, we can take that of visible light:

$$\lambda = 600 \, m\mu = 600 \times 10^{-7} \text{ cm}$$

$$m_f = \frac{6.62 \times 10^{-34} \text{ joule/sec}}{600 \times 10^{-7} \text{ cm} \times 3 \times 10^{10} \text{ cm}} = 3.68 \times 10^{-33} \text{ grams}$$

$$\frac{m_e}{m_f} = \frac{9.11 \times 10^{-28}}{3.68 \times 10^{-33}} = 2.47 \times 10^5 = 247,000.$$

The Energy Ratio

$$E_e = m \, c^2 = 0.51 \times 10^7 \text{ joule/sec}$$

$$E_f = 3.68 \times 10^{-33} \times 3 \times 10^{10} \text{ grams} \frac{\text{cm}}{\text{sec}^2} = 11 \times 10^{-23} \frac{\text{dina}}{\text{sec}}$$

$$= 11 \times 10^{-30} \frac{\text{joule}}{\text{sec}}$$

10

$$\frac{E_e}{E_f} = \frac{0.51 \times 10^7 \text{ joule}}{11 \times 10^{-30} \text{ joule}} = 4.6 \times 10^{39}$$

The Distance Ratio

The diameter, D_H, of the orbit of the hydrogen-atom electron is 0.53×10^{-8} cm, and the wavelength taken for light in the visible spectrum, λ, as above, is 0.6×10^{-4} cm; thus we have

$$\frac{D_H}{\lambda_\gamma} = \frac{5.3 \times 10^{-9}}{6 \times 10^{-5}} = 0.8 \times 10^{-4}.$$

Then $\lambda \cong 10{,}000 \, D_H$.

If we take the wavelength of gamma rays (λ_γ) to be 10^{-12} cm, we may write

$$\frac{D_H}{\lambda_\gamma} = \frac{5.3 \times 10^{-9}}{10^{-12}} = 5.3 \times 10^3.$$

Thus $D_H = 5000 \, \lambda_\gamma$ (see Figure 4).

$$\frac{t}{\Gamma} = \frac{0.5 \times 10^{-8} \times 3 \times 10^{10}}{600 \times 10^{-7} \times 2\pi} = 18.84 \times 10^{11}$$

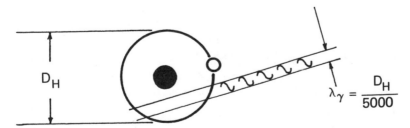

$$\lambda_\gamma = \frac{D_H}{5000}$$

Figure 4

Thus, summarizing, we have the following ratios and values:

$$\frac{m_e}{m_f} = 247,000$$

$$\frac{E_e}{E_f} = 4.6 \times 10^{39}$$

$$\lambda = 10,000 \, D_H$$

$$\lambda_\gamma = \frac{D_H}{5,000}.$$

And we note that the mass ratio is also of the order of tens of thousands.

The energy ratio does not concern us directly in the present matter, but it does serve to show the low level of the energy of the photon and hence the reason why it is virtually unaffected by potential fields, its long life, and the supposed constancy of its velocity.

The relations between the visible-light and gamma-ray photon wavelengths and the diameter of the hydrogen-atom electron orbit are an indication of its wide range, running from thousands of times smaller than the electron orbit to hundreds of thousands of times larger, as would be the case for radio waves.

This gives rise to the supposition that the wavetrain or packet is the result of a series of corpuscles, in the ee domain, in association with others revolving about them giving rise to some perturbation in the potential fields through which the wavetrain may pass.

In the same fashion that the proton is associated with the electron, we may refer to the photon being associated with a particle to be known as the *quasitron*. For e/ee time calculations then, we can take the period of the electron in its orbit and that of the quasitron, based on the wavelength of visible light.

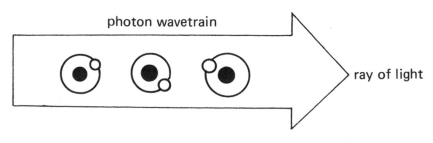

Figure 5

12

Thus, we have the domain passages as follows.

Domain U

$$\frac{T}{t} = 6 \times 10^{21} \qquad \frac{V}{v} = 1 \qquad \frac{M}{m} = 0.54 \times 10^{57} \qquad \frac{L}{l} = 0.56 \times 10^{22}$$

$$\frac{t}{T} = 1.6 \times 10^{-22} \qquad \frac{v}{V} = 1 \qquad \frac{m}{M} = 1.52 \times 10^{-55} \qquad \frac{l}{L} = 1.8 \times 10^{-22}$$

Domain e

$$\frac{t}{\Gamma} = 1.8 \times 10^{12} \qquad \frac{v}{\nu} = 1* \qquad \frac{m}{\mu} = 2.4 \times 10^{5} \qquad \frac{l}{\lambda} = 2 \times 10^{-4}$$

$$\frac{\Gamma}{t} = 1.5 \times 10^{-6} \qquad \frac{\nu}{v} = 1 \qquad \frac{\mu}{m} = 4 \times 10^{-6} \qquad \frac{\lambda}{l} = 0.5 \times 10^{4}$$

Domain ee

Some conclusions: From the little data available, we can arrive at an approximation of the changes of scale between domains. We see that in a shift from U to e there is an extremely large change in time, mass, and distance, but not in velocities.

In a shift from e to ee the scale change for time is still very large but not so large as in the previous case. As for mass and distance, the scale change is relatively small in comparison with the other values.

In shifts from U to e the ratio of the scales of distances to those of time should be equal to the velocities scale, and, in fact, we find (fortunately) that the value is approximately unity (represented by the number 1).

If we take a shift from e to ee, however, such a ratio is simply not good enough. This, of itself, is not surprising given the values taken, but, even so, it must be improved until the results coincide. If this coincidence were not to occur, of course, the principles of conservation of energy, mass and momentum between domains would be violated.

*See Chapter III.

13

Chapter II

THE NATURE OF GRAVITY

BODE'S LAW

From Newton's equation, we know that the gravitational attraction between two bodies is given by an inverse square relationship (spherical diffusion) and is proportional to the "product" of the masses of the bodies.

We shall go on to see the reasoning process whereby we arrive at current concepts put forward to explain the nature of gravity. We shall also consider the nature of the equations of Newton and Coulomb to see whether the unification of the fields of gravity and electromagnetism is, in fact, a requirement for the progress of science or, rather, a simplification of matters arising from a wish for a unifying concept on our part, with little correspondence to the realities of nature.

The product of the masses M and m shows that each element of M acts on each one of m, and vice versa.

If a hole were to be drilled to the center of the earth, the gravity value at the center would be zero. A body falling freely downward would initially be subject to an acceleration, subsequently to a deceleration, and would finally arrive at a state of rest at the center.

Newton's expression alone describes a two-body system well enough but is not satisfactory for a planetary system. The inclusion of more than two bodies implies an effort to arrive at a total field of force generated by the masses, but there is no corresponding solution for a complete description of any such system. The most significant shortcoming is in the lack of justification of Bode's Law. This is the geometric series

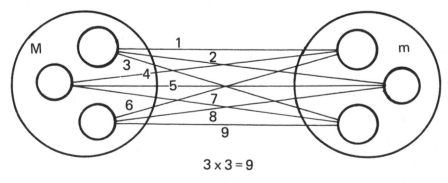

$3 \times 3 = 9$

Figure 6

14

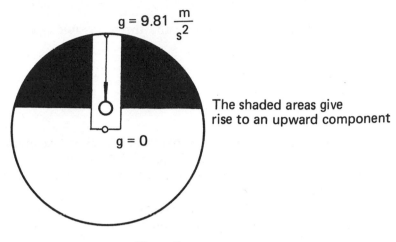

$$g = 9.81 \frac{m}{s^2}$$

$$g = 0$$

The shaded areas give
rise to an upward component

Figure 7

giving empirical values for the distances of the planets from the sun, namely, 0-3-6-12-24-48-96-192-384-768. Since it is reasonable to suspect that an energy term is what is missing, we can consider the potential candidates.

The options would be centripetal force, kinetic energy, and momentum. We can dismiss both centripetal force—since we need rotation with two bound masses—and kinetic energy—because we are aware that it is the integral of a force and bears upon energy accummulation, thrust, and suchlike; we are left with momentum, the product of mass and velocity.

We can go through a sequence of deduction as follows: In a binary system, we know from Newton that the maximum attractive force of the sun would be with a body similar to it. If we take $F*$ as the maximum total force of the binary system, and N is to represent other bodies in the system, with F as the attraction with respect to the sun, we have the following (where $G = 6.7 \times 10^{-8}$ cm^3/g sec^2):

$$F* = G \frac{2M}{d^2}$$

$$F* > \sum_1^n F N_i.$$

Thus the latter equation represents an unsaturated planetary system. To show this is the case, we can proceed as follows:

15

$$\boxed{Cms(\text{sun}) \lessgtr \Sigma\, Cmp(\text{planets})}$$.

Where Cm is the momentum; the three possible cases would then be as follows:

$Cms = \Sigma\, Cmp$ Saturation and equilibrium

$Cms > \Sigma\, Cmp$ Unsaturated

$Cms < \Sigma\, Cmp$

In the last case, the process is inverted, a kind of transition takes place, and a new equilibrium is established. The body having a Cm greater than the sum of those of the others becomes the center, or the planets unite until this becomes the case.

Movements in which the sun drags planets in or the system moves as a whole are not considered. The equation for the equilibrium of momentum is as follows:

$$M_0 (V_{R_0} + V_{T_0}) = m_1 (v_{R_1} + v_{T_1}) + m_2 (v_{R_2} + v_{T_2}) +$$
$$\cdots + m_{10} (v_{R_{10}} + v_{T_{10}})$$

   ;

the subindex 0 indicates the sun and the other numbered subindices the planets. In this expression, there is no correction for satellites, eccentricity of orbits and suchlike.

We shall use the data from the chart in Figure 11; V_R is the speed of a body's rotation about itself and v_T is the translation velocity in its orbit.

In the rotation (see Figure 8), it turns out that we can take the distance $r/2 = D/4$ as the distance at which all the mass may be centered for the purpose of momentum calculation. In other words, the calculation diameter in the reduced circumference is equal to the radius of the circumference of the body itself.

The Sun:

1) $M_0 = 332,000$ (Earth = 1)

2) $t_{R_0} = 25.35$ d, then $V_{R_0} = \dfrac{1}{t} = \dfrac{r/2\,\pi}{25.35} = \dfrac{696.000 \times \pi}{2,190,240} = 0.99$ Km/s

3) $V_{T_0} = 250$ Km/s

16

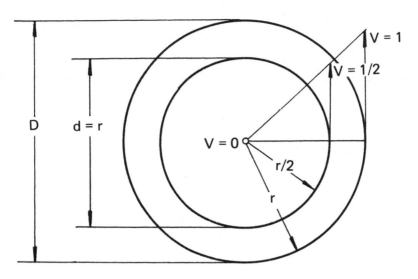

Figure 8

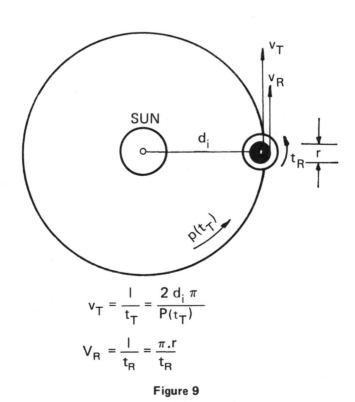

$$v_T = \frac{l}{t_T} = \frac{2\,d_i\,\pi}{P(t_T)}$$

$$V_R = \frac{l}{t_R} = \frac{\pi.r}{t_R}$$

Figure 9

Mercury: 1) Distance from the sun, $d_1 = 0.39$ (where earth = 1)

Major Semi axis.

2) $m_1 = 0.055$

3) $t_{R_1} = 58.6$ d, then $v_{R_1} = \dfrac{2{,}420 \times \pi}{5{,}063{,}040} = 1.5 \times 10^{-3} = 0.0015$ Km/s.

4) $t_{T_1} = 88$ d, then $v_{T_1} = \dfrac{0.39 \times 1.5 \times 6.28 \times 10^8}{7{,}603{,}200} = 48.3191$ Km/s.

Venus:

1) $d_2 = 0.72$
2) $m_2 = 0.82$

3) $t_{R_2} = 244$ d, then $v_{R_2} = \dfrac{6{,}100 \times 3.14}{244} = 0.00091$ km/s.

4) $t_{T_2} = 244$ d, then $v_{T_2} = \dfrac{1.08 \times 6.28 \times 10^8}{19.35 \times 10^6} = 35.05$ Km/s.

Earth:

1) $d_3 = 1$

2) $t_{R_3} = 23h56m4s$, then $v_{R_3} = \dfrac{6{,}378 \times 3.14}{1{,}990{,}564} = 0.01$ Km/s.

3) $t_{T_3} = 365.25$ d, then $v_{T_3} = \dfrac{1.5 \times 6.28 \times 10^8}{31{,}557{,}000} = 29.847$ Km/s.

Mars:

1) $d_4 = 1.52$
2) $m_4 = 0.11$

3) $t_{R_4} = 24h37m22s$, then $v_{R_4} = \dfrac{3{,}380 \times 3.14}{2{,}075{,}842} = 0.005112$ Km/s.

4) $t_{T_4} = 686.98$ d, then $v_{T_4} = \dfrac{1.52 \times 1.5 \times 6.28 \times 10^8}{5{,}935 \times 10^4} = 24.1253$ Km/s.

Asteroids:

1) $d_5 = 2.65$

Jupiter:

1) d_6 = 5.2

2) m_6 = 318

3) t_{R_6} = 9h50m, then $v_{R_6} = \dfrac{71,350 \times 3.14}{35,400 \times 4 \times 7} = 0.22$ Km/s.

4) t_{T_6} = 11.86 y, then $v_{T_6} = \dfrac{5.2 \times 1.5 \times 6.28 \times 10^8}{0.3156 \times 11.86 \times 10^8} = 13.0868$ Km/s.

Saturn:

1) d_7 = 9.54

2) m_7 = 95

3) t_{R_7} = 10h40m, then $v_{R_7} = \dfrac{60,400 \times 3.14}{38,400 \times 4 \times 8} = 0.15$ Km/s.

4) t_{T_7} = 29.45 y, then $v_{T_7} = \dfrac{9.54 \times 1.5 \times 10^8 \times 6.28}{29.45 \times 0.31 \times 10^8} = 9.84356$ Km/s.

Uranus:

1) d_8 = 19.19

2) m_8 = 15

3) t_{R_8} = 10h 49m, then $v_{R_8} = \dfrac{23,800 \times 3.14}{38,940 \times 4 \times 8} = 0.05$ Km/s.

4) t_{T_8} = 84.01 y, then $v_{T_8} = \dfrac{19.19 \times 1.5 \times 6.28 \times 10^8}{84.01 \times 0.31 \times 10^8} = 6.9411$ Km/s.

Neptune:

1) d_9 = 30.07

2) m_9 = 17

3) t_{R_9} = 16 h, then $v_{R_9} = \dfrac{22,200 \times 3.14}{57,600 \times 4 \times 7} = 0.04$ Km/s.

4) t_{T_9} = 164.79 y, then $v_{T_9} = \dfrac{30.07 \times 1.5 \times 6.28 \times 10^8}{51.08 \times 10^8} = 5.5454$ Km/s.

Pluto:

1) d_{10} = 39.52

2) m_{10} = 0.1428

3) $t_{R_{10}}$ = 6.39 d, then $v_{R_{10}}$ = $\dfrac{2{,}750 \times 3.14}{552{,}096 \times 8}$ = 0.001 Km/s.

4) $t_{T_{10}}$ = 248.4 y, then $v_{T_{10}}$ = $\dfrac{39.52 \times 1.5 \times 10^8 \times 6.28}{248.4 \times 10^8 \times 0.31}$ = 4.7569 Km/s.

For distance calculations we employ $d_i = \dfrac{V_T \, P}{2 \times 3.14}$

The d_i values should be those of the Bode series, or, rather, the measured values. In the case of the earth, this would be:

$$d_3 = \frac{1.5 \times 2 \times 3.14 \times 10^8 / 365 \text{ d.} \times 365 \text{ d}}{6.28} = 1.5 \times 10^8 \text{ Km.}$$

It would also be possible to set up a ten-equation system to solve for the d_i values since $V_T = (6.28 \times d)/P$, where P is the period.

$$M_o \, (V_{R_o} + V_{T_o}) = \overset{10}{\underset{1}{\Sigma}} \, m_i \, (v_{R_i} + \frac{2 \pi d_i}{P_i})$$

$$d_i = (M_o \, \frac{(V_{R_o} + V_{T_o})}{m_i} - V_{R_i}) \, \frac{P_i}{2 \pi}$$

A further consideration would be a system of equations for the whole planetary system, using force values corrected for the presence of the other bodies when Newton's Law for two is applied. In such a system, each distance would be a function of all the other masses and velocities. For example, a planet with a satellite would constitute a binary system having its own baricenter, with the bodies themselves virtually sharing the same orbit about the sun. The satellite in this case has sufficient momentum to maintain its distances both from the planet and the sun. If we ignore the rest of the bodies for a moment, such a system is one of three bodies in equilibrium.

If we suppose a moon (see Figure 18) that passes close to the solar system has sufficient momentum to keep it at a given distance from the sun, and at that distance there is a planet, then if the momenta are equal the bodies will have the same orbit. Further, if the velocity factors are the same, so too will be the masses, and they could go on, equidistant from the sun and each other, indefinitely in the same orbit. Should there be even a minute mass difference between the bodies, however, at equal rotational speeds, the translational velocity must change such that the bodies become closer. Finally, depending on the dynamics and their momenta, the bodies will either form a binary system of the earth-moon type—a single rotating body—or they may collide. These various options would allow for an explanation of captive moons, the rings of Saturn, asteroids, meteorites, and similar phenomena.

We could go further and formulate an Exclusion principle for a planetary system, for example, in the following terms: "In any stable system of bodies rotating in space, orbits are only occupied by a single principal body, unless there shall be two equal bodies forming a binary or planetary system. The location of the bodies is a function of the bodies formed in the system or later added to it or the momentum."

The orbital velocity of bodies further away from the sun is lower than that of those closer to it because the influence of the sun's gravity is correspondingly less at the greater distances. Even so, in order for a body to move to an orbit of greater potential energy—namely, an orbit further from the sun—it would have to somehow gain in rotational and/or translational velocity to arrive at the new orbit where the new values would be those dictated by the laws of Kepler and Newton.

Bode's law, which is empirical, is confusing since the distance coincidences it shows as following a particular sequence really only underlines that the sequence corresponds to the situation of the majority of the planets in the very special case of the stable configuration of the solar system. Bode's law has had its uses, but there is no fundamental reason to assume that the values it gives are unique, or the only ones possible. Provided the rotating bodies fulfill the requirements of equalization of forces, any orbit is possible. We can readily conceive of a system (see Figure 10) that does not comply with Bode's law.

In this case, for equal rotational velocities, the translational velocities would be in a sequence thus:

$$V_{T_1} > V_{T_2} > V_{T_3}$$

and the distance ratios would be 1:2:3.

Thus we can bring together the parameters applicable in U and e—that is, when a Bohr atom incorporates energy, an electron shifts to a higher orbit, and on dropping back closer to the nucleus, emits the same energy.

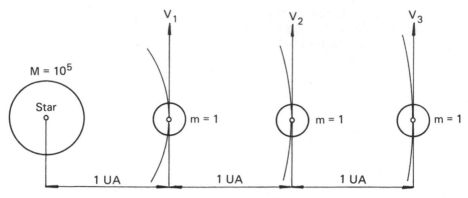

UA: astronomical unit (93,000 miles)

Figure 10

The following is a key to the table in Figure 11 (Solar Systems).

Row 1: Bode series
Row 2: Conversion based on earth-sun distance = 1
Row 3: Radii of the bodies
Row 4: Volumes
Row 5: Masses of bodies relative to the earth
Row 6: Densities
Row 7: Translational times
Row 8: Rotational times
Row 9: Number of Satellites
Row 10: Rotational Velocities
Row 11: Translational Velocities
Row 12: The sum of 10 and 11
Row 13: The product of the sum of the velocities by the mass
Row 14: The values we have calculated
Row 15: The direct or indirect reference measurements
Row 16: The difference between 14 and 15

The distances from the sun were calculated as follows:

$$V_{Total} = \frac{6.28 \times d_i}{P}, \text{ whence } d_i = \frac{V_{Total} \times P}{6.28}$$

where V_{Total} is the sum of the rotational and translational velocities.

22

			1	2	3	4	5	6	7	8	9	10
	Body No.											
	Name / Column of	SUN	Mercury	Venus	Earth	Marte	Asteroides	Jupiter	Saturn	Urano	Neptun	Pluton
1	BODE	—	0	3	6	12	24	48	96	192	384	768
2	+4&÷10	—	0,4	0,7	1	1,6	2,8	5,2	10	19,6	38,8	77,2
3	Radii Km	696.000	2420	6100	6378	3380	—	71350	60400	23800	22200	2750
4	Volume Earth = 1	—	0,055	0,88	1	0,15	—	1318	769	50	42	—
5	Mass Earth = 1	332.000	0,05	0,82	1	0,11	—	318	95	15	17	0,143
6	Density g/cm³	1,4	5,4	5,1	5,52	3,97	—	1,33	0,68	1,60	2,25	—
7	Translation	—	88 d.	224 d.	365,25 d	686,98 d	—	11,86 y	29,45 y	84,01 y	164,79 y	248,4 y
8	Rotation	—	58,6 d	244 d.	23 h 56 m 5 s	24 h 37 m 22 s	—	9 h 50 m	10 h 40 m	10.h 49 m	16 h	6,39 d.
9	Satellite	—	—	—	1	2	—	12	10	5	2	—
10	V Rot Km/Seg.	0,100	0,001	0,001	0,010	0,005	—	*0,26	*0,24	*0,07	*0,005	0,015
11	VT Km/Seg.	250	48,32	35,05	29,85	24,12	—	13,08	9,84	6,94	5,54	4,83
12	VR + VT KM/Seg.	250,1	48,321	35,051	29,86	24,125	—	13,34	10,08	7,01	5,59	4,845
13	(VR + VT)m	$8,3 \times 10^7$	2,41	28,74	29,86	2,65	—	4325	974	106	96	0,69
14	Calculated	—	0,39	0,72	1	1,52	—	5,30	9,85	19,67	30,25	39,52
15	Measured	—	0,39	0,72	1	1,52	2,65	5,22	9,54	19,19	30,07	39,52
16	Difference	—	0,00	0,00	0,00	0,00	—	0,08	0,31	0,48	0,18	0,00

Figure 11

*Adopted

In this fashion, we arrive at the following values:

$$d_1 = \frac{48.3206 \times 7.6}{942} = 0.3899$$

$$d_2 = \frac{35.0519 \times 19.35}{942} = 0.7199$$

$$d_3 = \frac{29.8570 \times 31.56}{942} = 1.0003$$

$$d_4 = \frac{24.125 \times 59.35}{942} = 1.5199$$

$$d_5 = \text{Asteroids}$$

$$d_6 = \frac{13.34 \times 374.30}{942} = 5.30$$

$$d_7 = \frac{10.08 \times 929.44}{942} = 9.85$$

$$d_8 = \frac{6.99 \times 2,651.35}{942} = 19.67$$

$$d_9 = \frac{5.58 \times 5,108}{942} = 30.25$$

$$d_{10} = \frac{4.7569 \times 7,826}{942} = 39.52.$$

For the large-diameter/low-density planets Jupiter and Saturn, along with Uranus and Neptune, we have introduced values for V_{Total} as given by consideration of the observations (2)(a) through (d) below, since this gives an agreement with the reference values. These same observations would be applicable to the Sun insofar as the value for its rotational velocity would not be definite.

If we take the aggregate momentum value of all the planets to be 5,565.35 and compare it with that of the Sun (8.3×10^7) we find a ratio of 1:14,913.70. Such a value would, of course, correspond to an unsaturated system.

The observations mentioned are as follows:

(1) We have arrived at the distance values by calculation, just as Kepler did, employing the data which required explanation. We should, however, bear in mind that there is a value—that of rotation—that is independent, and hence from this value we can endeavor to unify unknowns with the facts we know in order to set up an appropriate formulation of the problem.

(2) The coincidences relative to the reference values can be considered good for the planets up to Mars since these are high-density planets

24

having almost no moons. From Jupiter outward we would make the following observations:

(a) The distance values are greater than the reference values and should be verified—first, because many of these reference measurements are not the result of direct observation or calculation and hence are at least approximate, if not openly hypothetical. An example would be the calculation of a rotational period of a mass from observations of a spot on its surface, which itself is virtually liquid or gaseous. Second, when the reference values signify greater mass or velocity contributions, then they need to be corrected.

(b) In the mass values, satellites and the rings of Saturn were not considered. The masses were also calculated using Newton's formula, something that has to be closely examined.

(c) Owing to friction in fluids and the low densities prevalent in these planets, the rotational-velocity values are hardly correct—that is, the mean rotational values should be greater in these cases.

(d) The influence of other large masses in the region, specifically Jupiter and Saturn, has now been properly taken into account.

(e) It is probable that these error factors, in varying degrees, influence matters, and that we can arrive at a correction factor K since we can write a relationship as follows:

$$d_i = \sqrt{G \frac{(M + m_i)}{F_i}} = K \frac{V_i P_i}{2\pi}.$$

The most important conclusion is that with a variation of the total velocity of each mass we have another equilibrium condition. It is total velocity since, if we reduce V_T and increase V_R at constant total velocity, then the body remains in its orbit even if one value were to fall to zero, as long as the other is correspondingly increased to maintain the original total velocity value.

In Figure 11, we observe that the rotational values of planets six to nine—Jupiter, Saturn, Uranus, and Neptune—are very low in comparison with the values given in astronomical tables. With such values, the distance from the sun turns out to be six astronomical units too large; in other words, this shows either a greater distance and hence a larger mass than that calculated, or a larger velocity. If we take the mass value as good, we can look at the physical consequences with these planets in connection with this discrepancy.

In Figure 12, we have the radial-density variation curve for the Sun, from the center out to limbo. Now, since we know the planets that concern us have a physical make-up similar to that of the Sun, we can,

25

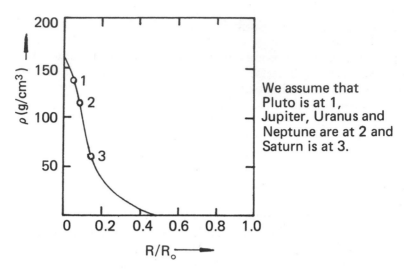

We assume that
Pluto is at 1,
Jupiter, Uranus and
Neptune are at 2 and
Saturn is at 3.

The Sun: Radial density variation from the center
to limbo. (*Source*: R.L. Sears, *Astrophysical
Journal*, vol. 140, p. 477, 1964)

Figure 12

at least provisionally, use the curve to see whether we can approach our unknown more closely.

In Figure 13 we have an estimation for the calculation radius of these low-density planets to replace the approach for the high-density cases. (Note: if a planet is of low density but small size, like Mercury, the distance anomalies with which we are concerned are not apparent.) Here, then, we see that three quarters of a planet's mass is concentrated inside one third of its diameter: the calculation radius should be one sixth of the overall diameter, and we have to consider one third of the remaining mass. We can round off to this value when estimating the rotational velocity in treating this phenomenon.

The fact is, the high-density nucleus rotates more rapidly than the surface. The astronomical data for surface velocity comes from direct observation, so we shall suppose that at one sixth of the diameter from the center, the rotational velocity is four to eight times the surface value. These are the values for these planets given in Figure 11.

It is interesting to note that if we were to require more precision by taking values proportional to the densities, errors would also be proportional and hence more balanced; such values are those presented in the tables.

In the case of Pluto, given the imprecision of the data because of the large distance involved, along with other problems inherent to the planet,

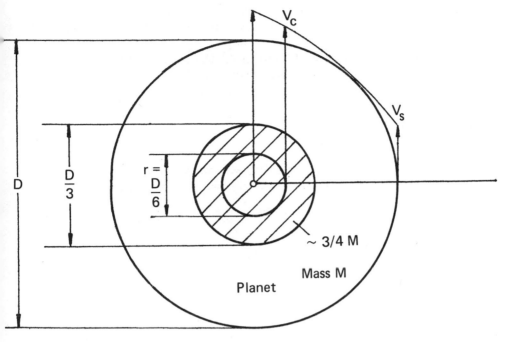

Estimation of the calculation radius of the low-density planets.

$$(4 \text{ to } 8) \ V_s = V_c \quad \text{where } V_s = \text{Surface velocity, and}$$
$$V_c = \text{Calculation velocity}$$

Figure 13

we have assumed that it conforms to the situation prevailing for the inner planets.

In this fashion, we have endeavored to show that Bode's law, while having served the purpose of stimulating research, is but a numerical coincidence, and even then it is only valid in the first nine terms of its series. This, of course, is not to say that systems may be found that conform to Bode's law in more terms than nine. This would serve as negative proof of Bode's law.

These calculations have also served, once the low-density planet discrepancy is incorporated, to make a correction for their physical nature. This has not only supported the criteria adopted but also allows us to see the lack of uniformity in planetary systems, even one as apparently simple as that of our Sun. We have also assembled a tool to determine, from a knowledge of the surface velocity of a planet or other celestial body rotating in a system in equilibrium, the velocity distribution across successive concentric rotational layers and the corresponding densities, all of which allows for some elucidation of the internal physics of such planets.

In the above, we have considered that the rotational baricenter correction effects for our own and other planet systems are minimal, as also are such corrections for satellites, eccentricy, *et cetera*.

NEWTON'S LAW

In light of the Bode's law results, we can consider Newton's law. If we take momentum (mass x velocity) instead of simply mass, we have

$$F = G\, \frac{M\, m\, V\, v}{d^2}\, , \qquad (1)$$

whence we see the force of attraction as being proportional not to the product of the masses but rather to that of the momenta. This has its attractions, since it is a dynamic process.

Thus this first equation would be a particular case of the general expression for a multibody system which we used when considering Bode's law, and at the same time Newton's law would be a particular case of that of Bode. For these cases, if the velocities were zero, then F would also be zero, whereby we would have masses at rest, and if the force between them were to be zero, the U domain would be in total equilibrium.

Let us consider what happens if we take the first equation instead of Newton's formula.

1) $V = 0$

    $v = 0$, then $F = 0$

2) $V = 1$

    $v = 1$, then $F = G\, \dfrac{Mm}{d^2}$

3) $V = 32$

    $v = 16$, then $F = -G\, \dfrac{M\, 32\, m\, 16}{d^2} = -G\, \dfrac{M\, \frac{32}{16}\, m\, \frac{16}{16}}{d^2} = -G\, \dfrac{2Mm}{d^2}$

In Figure 14, we see the following cases:

a) $\overline{V} = \overline{v} = 0$; $F = 0$ and the bodies are in contact. (Note: the line over the velocities indicates total velocity, i.e. rotational translational)

28

(a) $F = 0$

$\dfrac{\overline{V}}{\overline{v}} = 0, \overline{V} = 0$

$F \equiv$ Quietude

$$M \equiv m$$

(b) $F = -G \dfrac{M.m. \, \overline{V}/\overline{v}}{d^2}$

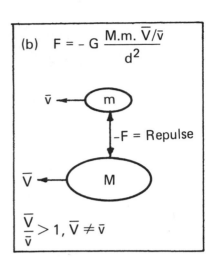

$\dfrac{\overline{V}}{\overline{v}} > 1, \overline{V} \neq \overline{v}$

(c) $F = +G \dfrac{M.m. \, \overline{V}/\overline{v}}{d^2}$

$0 < \dfrac{\overline{V}}{\overline{v}} < 1; \overline{V} \neq \overline{v}$

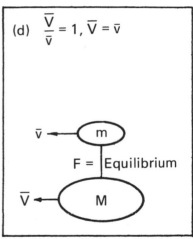

(d) $\dfrac{\overline{V}}{\overline{v}} = 1, \overline{V} = \overline{v}$

Newton's, Kepler's, etc. laws

Figure 14

b) In this case, we have: $\overline{V} > \overline{v}$; $\dfrac{\overline{V}}{\overline{v}} > 1$; $F = -G \dfrac{M.m. \, \overline{V}/\overline{v}}{d^2}$

i.e.: $F = -G \dfrac{1,3Mm}{d^2}$

and the bodies move apart.

c) In this case, we have $\overline{V} < \overline{v}$; $\dfrac{\overline{V}}{\overline{v}} < 1$; $F = +G\,\dfrac{Mm\overline{V}/\overline{v}}{d^2}$

i.e.: $F = +G\,\dfrac{0.6\,Mm}{d^2}$

and the bodies move toward each other.

d) In this case, we have: $\overline{V} = \overline{v}$; $F = G\,\dfrac{Mm\overline{V}/\overline{v}}{d^2}$; $\dfrac{\overline{V}}{\overline{v}} = 1$

i.e.: $F = G\,\dfrac{1.M.m}{d^2}$

and we have equilibrium.

The ratio $\overline{V}/\overline{v}$ has no negative values.

We can now proceed to analyze the escape velocity of rockets, Cavendish's balance, and the relativistic expressions for gravitational fields in this context. We shall take a rocket, in escape, where

$$V = 11.2 \text{ km/sec and applying } F = G\,\frac{M.m\,\overline{V}/\overline{v}}{d^2} \qquad (2)$$

we have, $\overline{V}/\overline{v} = \dfrac{30\ Km/sec}{11.2\ Km/sec} = 2.67$ and

$$F = -G\,\frac{M.m.\,2.67}{d^2} = -G\,2.67\,\frac{M.m}{d^2}$$

There is no control applicable with $F = \dfrac{G.Mm}{d^2}$.

This means that the force of attraction has been reduced by a factor of 2.67 relative to that in the case of the two bodies traveling through space at the same velocity, where this "same velocity" is understood as the sum of all the velocities of the masses.

In the case of Cavendish's balance, since everything is at rest on the surface of the earth then the bodies have the same rotational and translational velocities—namely, those of the earth itself—thus the ratio $\overline{V}/\overline{v} = 1$.

Equation (1) becomes

$$F = G\,\frac{M\,1\,m}{d^2} \ .$$

This result remains unchanged with the relativistic formulae, since these consider the gravitational field for phenomena occurring in the field and in which the impulse quadrivector remains constant.

30

We cannot apply Equation (2) for the sun-earth system since it is an unsaturated planetary system. If it were saturated and the value $\overline{V}$ were known, the both Equation (2) would apply and the condition $\overline{V}/\overline{v} = 1$ would be fulfilled.

There would also be fulfillment as in some of the cases in Figure 14.

The momentum value $m_i v_i$ represents Newton's centripetal force, a, where $a = v^2/d$, and was the force that prevented the moon from falling in toward the earth. These forces are not the same, insofar as we believe that, in the absence of a physical bond like a cord linking a rotating stone to a center of rotation, the centripetal force similarity should be abandoned.

In Figure 15 we see a sketch of the following situation: An automobile approaches a bend at excessive speed, as the tires begin to lose adhesion and at some point they do so; the wheels then actually leave the road surface. The sudden separation allows the release of the energy originally

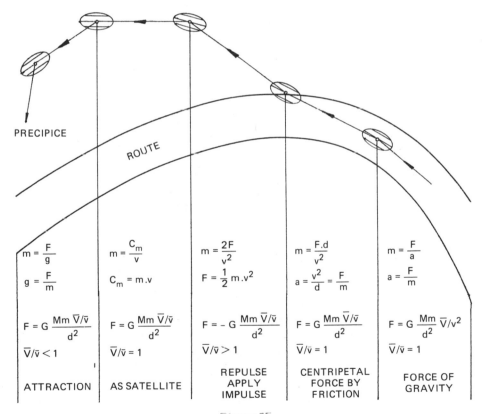

PRECIPICE

ROUTE

$m = \dfrac{F}{g}$	$m = \dfrac{C_m}{v}$	$m = \dfrac{2F}{v^2}$	$m = \dfrac{F.d}{v^2}$	$m = \dfrac{F}{a}$
$g = \dfrac{F}{m}$	$C_m = m.v$	$F = \dfrac{1}{2}m.v^2$	$a = \dfrac{v^2}{d} = \dfrac{F}{m}$	$a = \dfrac{F}{m}$
$F = G\,\dfrac{Mm\,\overline{V}/\overline{v}}{d^2}$	$F = G\,\dfrac{Mm\,\overline{V}/\overline{v}}{d^2}$	$F = -\,G\,\dfrac{Mm\,\overline{V}/\overline{v}}{d^2}$	$F = G\,\dfrac{Mm\,\overline{V}/\overline{v}}{d^2}$	$F = G\,\dfrac{Mm}{d^2}\,\overline{V}/v^2$
$\overline{V}/\overline{v} < 1$	$\overline{V}/\overline{v} = 1$	$\overline{V}/\overline{v} > 1$	$\overline{V}/\overline{v} = 1$	$\overline{V}/\overline{v} = 1$
ATTRACTION	AS SATELLITE	REPULSE APPLY IMPULSE	CENTRIPETAL FORCE BY FRICTION	FORCE OF GRAVITY

Figure 15

31

expended in the abrasive adhesion/skid (thus, impulse or thrust I = Δ Cm = m V_2 - mV_1) and the automobile rises slightly, traveling virtually parallel to the road surface, only to leave the road and initiate a fall over a steep incline. In the figure, different formulae are seen to be applicable for each stage of the incident.

In order to support the above concepts we shall pass through the domains, from U to e and on to ee.

We shall take the conditions for equilibrium considering gravitational forces alone and broaden the application thereof to a "planetary system."

At present, the limits of our knowledge are insufficient for a proper determination of the nature of gravity. We must move down to e and to ee, where we have to distinguish what happens to the elements of these domains so that, back in U, we may classify the nature phenomena as gravitational, electrical, magnetic, or electromagnetic. In Figure 16, we have diagramatic representations of such cases.

In the case of gravity, masses exert a mutual attraction because, in an atom, the sum of the excess capacity for equilibrium between nucleus and rotating elements in orbit gives rise to an overall force in the total mass. This force is small in comparison with electric and magnetic cases, and we shall see below how it is propagated through space. First, however, we shall compare the electrostatic repulsion between two alpha particles separated by a distance of 10^{-11} cm (9.18 x 10^{-2} Newtons) with the gravitational attraction between the particles being 2.97 x 10^{-37} Newtons: the resultant force is one of repulsion, equal to 6.21 x 10^{-35} Newtons. Now, in Figure 16, case (a) is an unsaturated planetary system; the remaining unsaturation is the value of the gravitational capacity contributed by the system.

The theory put forward here is sufficient to allow calculations for the construction of an antigravity craft as shown diagramatically in Figure 19. Such a craft would have a performance two or three times beyond that of conventional aircraft.

In the solar system itself, if it were possible to increase the rotation rate of all the planets, they would move to higher potential-energy orbits farther away from the sun, converging at a distance such that the whole would finally become a binary system, Sun-other mass.

In the particular case of the masses being equal, the sum of their individual rotations and translations would also be the same. If such a system were in a saturated equilibrium, this would not mean that gravity would be cancelled out, but that, since the system would have mass and velocity, it would also have gravity and hence would consititue the construction of the U domain.

In Figure 17, we see how the surface friction between mass 1 and mass 2 gives rise to an energy transfer from the electrons A to the velo-

(a)   Domain e or ee

Gravity in U:
Not excessive rotation and translation
of masses, equilibrium of attractive
and repulsive forces. Note that we are
in domain e or ee.

(b)   Domain e

Electric Charges in U:
Excessive rotation and translation of
masses by external forces aggregated.
Large repulsive forces.

(c)   Domain e

Magnetism in U:
The same as (b), but with small
repulsive forces.

(d)   Domain e or ee

ElectroMagnetism in U:
Rotation and translation of masses,
excessive translation of nucleus,
equilibrium of attractive and
repulsive forces.
Note that we are in domain e or ee.

Figure 16

city of the electrons B. If the friction is continued, the result of a thrust caused from the incremental potential field in B_1 toward B_2, from B_2 to B_3, and so on, thus giving a displacement of the incremental potential fields in e at the surface of the metal and possibly down through several atomic diameters, giving rise to the electric current V.

This example enables us to establish a similarity between events in U, gravitational phenomena and events in e (also of the planetary and gravitational type). From U these appear as electrical in nature, since the increase in the movements from the B elements by the action of A elements gives rise to an electric current.

The hidden interpretation factor must be clear and, we believe, accessible to us if we accept the criteria described.

We have seen that gravitational potential "between the bodies" may cancel out, for example, in a binary system, but, as a whole, these elements continue to be a mass with gravitational capacity. There is nothing to prevent us supposing that two systems can rotate about each other. In fact, we see just such in the successive increments of planetary systems with stars up to galaxies, with stars in the role of planets.

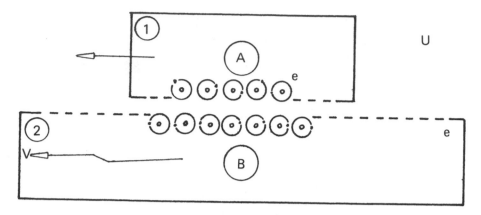

Figure 17

So it is that nature works—dynamics arise from the incorporation of movement and consequent exertion of force upon masses. We are not aware of whether, in the final analysis, we can reduce everything to forces, but we do know that, in their absence, the masses would form a single compact unit in space, leaving no separation between its constituent parts.

If we were to move down to domains e and ee on the basis of the foregoing criterion, we would find that in e, with its planetary systems in conditions of equilibrium, there is an attraction between masses; indeed, such forces must have been those that first brought it to equilibrium. In e, as in U, gravity does not disappear; rather, as we saw in Figure 16, it can, depending on the value of that energy, be seen from U as a mass of simple gravitational, electric, or similar capacity.

We may fruitfully ask ouselves about the gravitational equivalent of repulsion that would be comparable to the repulsion found between like electric charges. The answer is straightforward: If sufficient energy is added to a mass in U, as occurs in e, a repulsion will be observed. In other words, if we were to add kinetic energy to the earth, it would move away from the sun, and on observing an increase of distance with time, we say that we are observing repulsion.

We should clarify whether, in space, a mass distorts the space or the distortion of the space is due to the presence of mass. Both these appear to be different ways of saying the same thing. This would bring us to a definition of space from ee through to UU as "that region empty of mass and without any dimension," insofar as dimensions only arise when there are masses. It would follow that a dimensionless space could not even be infinite.

34

From the above we can see that the distortion of space in the presence of masses in movement is an interference between quasispherical fields; it is heterogeneous, and there exists a certain field density depending on the zone (see Figure 18).

The limits of the universe are, because of their heterogeneity, undefinable; however, if it were possible to "see" U from UU, its shape could be established.

If U were homogeneous there would be no movement; the observed movement of masses tends toward a homogeneous distribution of matter in space.

Bodies in space move from high field-density zones to ones of lower field density. In the shaded zone in Figure 18, showing the region of interference between local fields, mass-displacement phenomena are more pronounced.

Now, in U, gravitational phenomena must be described in spherical relativistic coordinates; hence if a mass distorts space in the region in which it moves, there would be new curvilinear coordinates in such regions. This supposition, while satisfying, in a way, is also sterile. Thus, if mass without energy was at some time concentrated, without force in U or e, or uniformly distributed in space without force in U but having force in e, then action over a distance would be a confusion of concepts. This can be approached by separating the terms.

In the example offered by our solar system, there is distance, but the action is not to be found at the Sun but at the planet. The planet has (opportunely acquired) energy, which allows it to remain at a certain distance from the Sun; thus action over a distance is not what is observed, since the action is to be found mechanically in evidence in the body of the planet. If this energy were somehow anulled the planet would fall in

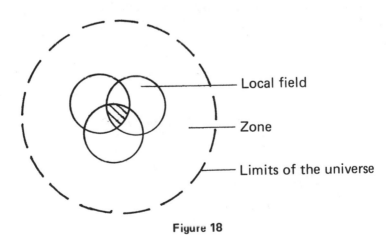

Figure 18

35

toward the Sun, hence increasing its mass. The process could be carried further: If the Sun were also to lose energy it would proceed to fall toward the center of the galaxy, and so it could go on to produce a kind of energy death of the universe.

It would appear that the spherical diffusion of gravity is not uniform insofar as the heaviest planets are located in one plane in which a greater amount of mass covers a larger distance, or, more to the point, where their product is greater.

Now, independent of the gravity constant g for earth, the hydrogen atom could be assigned a mass of unity. On this basis, a "periodic table" of the gravitational capacity of a gram of each element could be constructed, and hence, by using Avogadro's number N, that of any mass.

Since gravity affects the path of light, it would be conceivable that gravitational phenomena are common to all domains, while light phenomena belong to the ee domain. In this context, the following expression is useful:

$$F = \pm k \, \frac{M_i . \overline{V}_i . m_i . \overline{v}_i}{d^2}$$

The reason why galaxies are moving away from one another at speeds of 100 to 1000 km/sec could be explained: By using Newton's formula we obtain a value for the total mass of our galaxy of 2×10^{11} times that of the Sun, while in reality the sum of the masses of its stars gives a value of half this figure. We do not know, should we wish to add it in for the numbers to coincide, whether the interstellar dust (3 atoms per cubic centimeter) is in repose or in movement.

The position of Mercury at its closest point to the sun (the *perihelion*) may also possibly be explained since the positions observed shift progressively from those predicted by 43 seconds of arc per century, and this has to coincide with its relativistic demonstration.

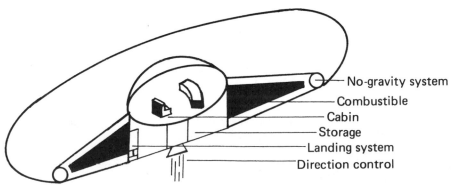

No-gravity system
Combustible
Cabin
Storage
Landing system
Direction control

Figure 19

36

In the universal system U, we note that at least half of the stars of one galaxy are binaries, double stars, or star systems of some higher multiplicity, rotating rapidly about one another. Of the simple stars, those having higher temperatures revolve on their own axes at enormous rates, while the cooler ones, such as our Sun, do so also, but more slowly. The explanation of this has to be that the stars of higher temperature, being younger, still have a relatively larger portion of their primordial energy, and hence a larger momentum. In our own solar system all the planets rotate in the same direction about the Sun, and their orbits are practically in the same plane; the angular momentum of the planets with respect to the Sun is also far larger than the Sun on its own axis. This is apparent from the equilibrium conditions of such an unsaturated system. It is also notable that the moons of the planets rotate in concentric circles, in the same direction and in planes close to that of the orbit of the planets.

THE NATURE OF GRAVITY

We have seen that we add energy, such as to increase the velocity of a body moving in a system in equilibrium, then that body will move outward. Now, if we consider the domain progression ee to e to U in regard to equilibrium actions equal for all of them, what is the action observed in the domain following the one in which we imagine ourselves to be?

First, let us bear in mind that they are not in equilibrium; then there is attraction in cases where some component does not have energy sufficient to maintain separation, and the attracting component has more. If we fail to grasp this, then we are constrained to return to the time of Newton, with action at a distance and the explanation that only God, who put things where they are, can understand them.

Let us look a little more closely at the matter as presented in Figure 20. In position 1 we see a concentrated, compact universe; in position 2, explosion and the energy then acquired by each part of the original mass; in position 3, the energy is distributed in space by the masses carrying it, and some degradation via primeval radioactivity occurs.

In position 4 we see in a part of U how masses M and M', having different energies owing to the unequal energy distribution in the initial explosion, have achieved equilibrium. We hardly expect there to be a lesser amount of energy per unit mass, but rather that there is a tendency to a uniform energy distribution in space and thus of the mass carrying it. We have assumed the "Big Bang" approach; other approaches, however, as far as the origin of gravity is concerned, bring us to the same conclusions.

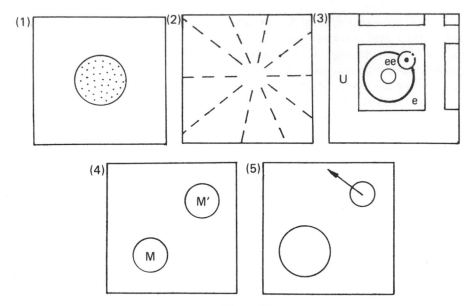

Figure 20

In position 5 we see the tendency of the masses toward the original compact state in the circumstances of less energy per unit volume, owing to space being unlimited, and a limited amount of mass with a constant amount of energy. This tendency we see in the attraction phenomena (action at a distance), which would appear to be the manifestation of what may be considered a process of recession.

Now let us look at details as in Figure 21.

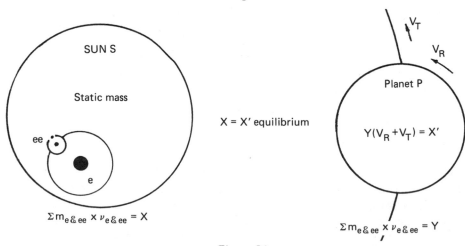

Figure 21

38

We can see that if X is equal to X′ there is an equilibrium where V_R plus V_T is added to V_e and V_{ee} such that $(V_R + V_T) = X'$.

If X′ is greater than X, the component bodies will move apart, but why? There is a repulsion because P moves to the region in space where its position of stability with reference to the baricenter of rotation of the binary system is located. In other words, there is a displacement to the position of equality of momentum, thus constituting a binary system, with P as the center and S rotating reciprocally.

If X′ is less than X, there is attraction, but why? The attraction arises because P, being short of energy but where the action is located, and there being no action at a distance, moves to the point in space where the other mass in the binary system is located with respect to the baricenter of the two—*viz*, it moves toward S. On arriving, it increases the mass of S, and the baricenter, which was moving as a function of the movement of P, is finally located at the center of S (see Figure 22).

In the case that X′ is greater than X, the body moves out to deep space, and so we see that masses to which kinetic energy is added behave in this fashion, increasing their potential energy with respect to the reference level. If this movement to deep space takes place, it is because the body does not have sufficient mass such that, with the increased velocity due to the kinetic energy added, it is able to destabilize S. The case with masses that are nearly equal is different; here, the lesser mass slowly attains the same momentum as the larger one, and the system becomes binary. It obviously follows that this latter mass, though the larger and originally the center, will begin to rotate. We can see the advantages of studying a binary system, with its two bodies in isolation, in order to arrive at an understanding both of gravity and systems having more than two bodies.

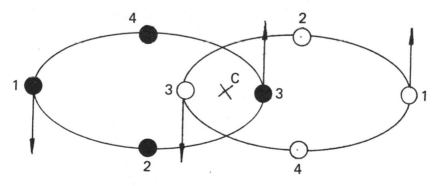

$\times$ C: Baricenter

Figure 22

39

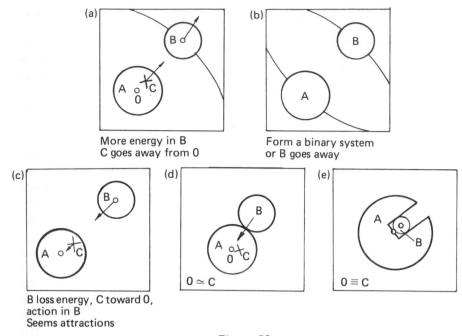

Figure 23

In Figure 23, we see the convergence of masses, movements of centers of rotation, and the movement of a mass whereby the attraction or repulsion of masses due to the gain or loss of energy (of the mass B) becomes apparent.

Case (c) in Figure 23 has given rise to confusion to date since it is but a part of the overall phenomenon. In the same way, in the solar system, rotation rates are so low that their significance has not been fully appreciated previously. We now see their importance, however, and it is possibly greater than that of translational velocity for high-energy-content systems.

The reader may, as an exercise, like to try explaining the tides along the lines here indicated, noting, however, that he must not use the phrase "the moon attracts," and if he is to resort to the use of mathematics, then to bear in mind that bodies of liquid, being fluids, are not rigidly attached to the earth.

In regard to the conditions prevailing in e or ee in conditions of supergravity—as would have been the case in the recessive process before the "Big Bang" and now is the case in white dwarfs and black holes and suchlike—we may, at least, infer the following: (1) It is a dynamic process. (2) On principle, energy is stored in ee. (3) The total energy E, in the domains, is limited just as the mass is, but both have a constant, determined value. In order, however, for a recession to occur and for there never to be an achievement of equilibrium with a fair distribution of the E in the component masses, we are constrained to believe that for such an equilibrium in that amount of mass, there is simply not enough E. (4) Such a situation may be reversed if, for example, further energy were to be received in U from UU.

This can be illustrated by a comparison with springs: The planetary system of ee is compressed upon itself by an increase in mass (with its momentum, Cm) and continues to do so "while it receives" mass (increase in gravity). The whole is kept under tension, but at the time it ceases to receive mass; the whole issue explodes and the expansion process begins again.

The planetary system of e or ee behaves as a compressible potential field with elastic characteristics. In the normal state we have a situation such as that represented diagramatically in Figure 24.

In the compressed or tensioned state, we would have the situation portrayed in Figure 25.

And here we note a change of sign! This means that the tensioned state for low-energy gravity conditions would explain antimatter and the changes of sign between fundamental particles. Black holes would constitute an example of such a process in particular regions of U. It is worthwhile considering the question of whether U reverts by regions or all together in one vast explosion.

Figure 26 is a representation of this process.

The successive processes of mass agglomeration give the tensioned equilibrium state, which, when the aggregation process ends, will explode.

The study of the signs must be made on the basis that we are handling tensioned states between potential fields.

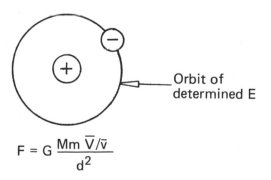

Orbit of determined E

$$F = G \frac{Mm \, \overline{V}/\overline{v}}{d^2}$$

Figure 24

41

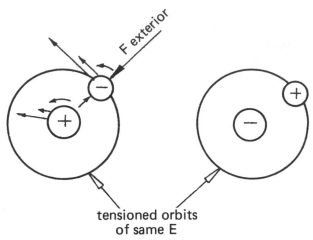

tensioned orbits
of same E

If f_{ext} predominates: is R the result of F and f_{ext}.

$$\boxed{F = G\frac{Mm\,\overline{V}/\overline{v}}{d^2} - f\ exterior}\quad \text{and with: } f\ exterior > G\frac{Mm\,\overline{V}/\overline{v}}{d^2}$$

$$\boxed{F = -G\frac{Mm\,\overline{V}/\overline{v}}{d^2}}\quad \text{and in high energy: } -F = C\frac{Mm\,\overline{V}/\overline{v}}{d^2}$$

Figure 25

I consider that the compact, concentrated state of all the material tensioned at one point does not necessarily have to be minuscule in size, as today is commonly supposed; nature herself seems to abhor infinitely small masses, infinitely high velocities, and other suchlike. We see, for instance, that in a drain situation if water overflows, or even in the subterranean flow of water through porous media, the water should reach an infinite velocity, and the water section should be a straight line. In practice, the velocity is finite and the water section is a surface. In the same line of thought, I think that the compact state mentioned above should have appreciable, though not very large, dimensions.

The achievement of supergravity in the laboratory would serve not only to clarify this matter, but could lead to the production of "clean" energy and explosives, free of any secondary radiations.

For the case of an antigravity (not negative-gravity) vehicle, if we consider that $M\overline{V} = m\overline{v}$ and hence $\overline{v} = M\overline{V}/m$, where $\overline{v}$ is velocity of the vehicle and $\overline{V}$ that of the earth, for the case of flight in the neighborhood of the earth, then we find extremely high values of $\overline{v}$ insofar as we have a binary system.

42

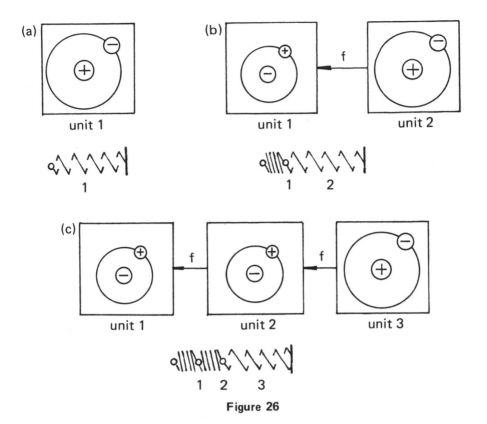

Figure 26

For the calculation of total loss of gravity we would consider $F = \pm G (Mm\overline{V}/\overline{v})/d^2$, and since we are not looking for F, but rather for the gravity-free condition, then we would have $\overline{V}/\overline{v} = 1$; and since $\overline{V} = 30$ km/sec, this condition would be fulfilled with a $\overline{v}$ value of 30 km/sec.

We should note that at the escape velocity, the ratio $\overline{V}/\overline{v}$ is equal to 2.67 and F is negative, with $\overline{v}$ equal to 11.2 km/sec. Now, if $\overline{v}$ were to be 15 km/sec, then the above ratio, $\overline{V}/\overline{v}$, would have the value 2: The higher the speed, the lower the value of the coefficient applicable to F in the journey of the vehicle to the equilibrium, or gravity-free position. With a $\overline{v}$ value of 30 km/sec, the ratio becomes unity, and this is the case of geostationary satellites.

A greater velocity of escape means a lower F value, going on to repulsion, since the process is dynamic. If we were to wish to have a value for F, we would need the mass values.

Now, the gravity-free state depends only upon the velocity and is independent of the masses involved:

43

$$\boxed{G = f(v)}.$$

It should, of course, be noted that the greater the mass involved, the more the energy to be added; the vehicle should therefore be as light as possible within the limits of the physical properties of the materials of which it is made. A plastic disc, for instance, rotating and moving through the air, requires less energy than would a metallic disc under the same flight conditions. If we should wish to take the lift available from the surrounding air for the metallic disc, then for given flight conditions we should consider rotation in a denser fluid than air—for example, water. Again, more energy input would be required in such a case.

Let us make some calculations in regard to an antigravity vehicle. Now, $\bar{v}$ must be 30 km/sec; with $V_T = 0$, then V_R would be 30,000 meters/sec.

If we consider a disc-shaped vehicle with mass uniformly concentrated at a distance from the center such that it moves 100 meters per revolution, then we would have the following:

$$\frac{30,000 \text{ m/sec}}{100 \text{ m}} = 300 \text{ rev/sec—}i.e., 18,000 \text{ rpm}.$$

The physical properties of steel are such that it would readily stand the tension forces resulting from such rates of revolution.

The $\bar{v}$ value has both V_T and V_R components, and these, in turn, are components of a dynamic process. For example, if the vehicle were to hover above the ground as does a helicopter, then the above calculation would apply.

If the vehicle were to move at 3,000 km/hr, then:

$$\frac{30 \text{ km/sec}}{3000 \text{ km/hr} + V_R} = 1$$
$$V_r = 30,000 \text{ m/sec} - \frac{3,000,000 \text{ m/sec}}{3,600}$$
$$= 30,000 - 833.33 \text{ m/sec} = 29.16 \text{ km/sec}.$$

For each 100-meter displacement of the center of mass we would have 291 rev/sec, equalling 17,460 rpm (for V_T value of 3,000 km/hr). In this case, the V_r value is more important than that of V_T.

In distance conditions such that the F value is small—for example, with a ratio $\bar{V}/\bar{v}$ value of 2—we would have the following:

44

$$\frac{\overline{V}}{\overline{v}} = 2.1, \text{ and thus } \frac{30 \text{ km/sec.}}{2} = 1;$$

hence $\overline{v} = 15$ km/sec.

The resulting rpm values would be smaller than before.

Among the factors that would have to be taken into account in the construction of a vehicle such as the one mentioned here, there would be (a) a contrarotating cabin; and (b) the shear effects of a disc on the surrounding medium (resistance of materials, lift-effect values for superheated air, cooling effect on air on the disc, flight beyond the atmosphere, conditions for stability, energy consumption, conditions of uniformly accelerated flight up to several Mach, and suchlike).

SUMMARY

At the beginning of time, there was a certain amount of mass in the universe, carrying a certain amount of energy in the form of velocity. The mass was dispersed, in many parts, as the result of a vast explosion within which we live. Although the universe is exploding, we do not normally see it as such because of the change in time scales; it happens before our eyes as though in very slow motion. Each portion of mass, then, has its own velocity.

When we look at the structure of the mass, we find it strange insofar as it is almost empty, and furthermore, a closer look reveals a spatial structure within this emptiness similar to that found outside of it. If this appears complex but acceptable, with some effort and in light of experimental evidence, what should our reaction be when, on looking even closer, we find in the inner particles a further repetition of the same structure? Having overcome our surprise at this apparently ongoing phenomenon, we readjust our understanding and draw the following conclusions.

Before the Big Bang, this inner, and very particular, spatial structure was under tension because of the enormous concentration of mass. When the pressure build-up of agglomeration came to an end, the whole exploded, like a spring on release from tension. Then, in the innermost parts, there was further tension, and in the explosion we observe this tension (energy) passing from inner parts to outer parts.

It should be realized that all energy-transfer phenomena are of this type—namely, from inner parts to outer parts. This would include solar energy passed to the earth, radioactive transformation, animal metabolism, and even clouds moving in the sky. Anything we observe forms part of this process, which, fundamentally, is the transfer of energy (velocities) from inner parts to outer ones.

45

The overall tendency is toward a possibly uniform distribution of the amount of available velocity in the totality of the mass available in the universe. When the process is complete we do not know what kind of situation will prevail. The fact is, we are also uncertain as to whether, in the process, there is any addition or subtraction of mass or velocity to or from the sum, both of which may take part in the event. There is also doubt regarding the initial Big Bang itself; it is, after all, but a postulate, which is acceptable for deducing the rest and explaining what we observe today.

Now, what happens with masses that give up velocity? Where do such masses come to rest? Let us consider the example of a dam. If we have water in a dam 1000 meters high, it has a potential energy of X. On release of the water, in a controlled manner through the turbines located below, its mass gives up energy and thereafter the water issues from the turbines and flows peacefully off down the riverbed at the base of the dam.

What happens in space? What constitutes the riverbed for a mass that gives up velocity? Where does such a mass encounter a lower point and equilibrium with the level of potential proper to itself? The answer would appear to be straightforward: We have already seen that in "near space" there is just such a potential field of equilibrium. An example would be that of the moon, on losing velocity and falling toward the earth; if the earth had a hole large enough to receive it, then the moon would come to rest at the "riverbed" of the center of the earth. The process, if repeated with the earth, would have it coming to rest at the center of the Sun, and so on to the center of the galaxy, et cetera.

Could such a situation give rise to a vast agglomeration, followed by a further "Big Bang"? We don't know; all we know is that velocities tend to distribute themselves. Thus bodies are not attracted by the earth but are subject to the tendency to distribute velocity and lose energy and so seek a lower point of rest.

Why does an object fall toward the earth? First, because it does not have sufficient velocity to remain at the level in which it is located. And second, in view of its lack of sufficient velocity, it seeks the energy "plane" proper to it unless prevented from so doing by some obstacle. The obstacle itself, though, is giving energy to the object by retaining it at some particular level. This, again, is a process that can be carried through with the obstacle of the obstacle, et cetera, at each stage describing the mechanics of equilibrium.

To follow through with this, it is necessary to go from domain to domain, and, in the observable universe, locate phenomena in the scheme of events. This is given to us by the laws of the tendency that those phenomena obey.

Chapter III

THE UNIFIED FIELD

COULOMB'S LAW

In Chapter I we saw a way of analyzing natural phenomena as a function of the domains. In a descriptive and mainly approximate fashion, we saw that with a change of scale to be achieved if we could reduce or increase our size enough, we could enter into the virtually unknown domains UU and ee, where we would encounter phenomena the same as those found in the more accessible domains U and e. In the light of the indications we now have at hand, we also saw that phenomena occurring in one domain, but observed and measured from another, sometimes take on somewhat disconcerting characteristics. Our example was that of gravitational phenomena in domain e which appear as electrical phenomena in domain U.

While these suppositions have a certain attractiveness, they may not entirely reflect reality. They are of use, however, if we apply them in an endeavor to clarify some of the more intriguing phenomena.

In Chapter II, we saw when looking at aspects of gravity that if we take Coulomb's law (where Q and q are charges)

$$F = -C \, \frac{Q \, q}{d^2}$$

and then make a domain shift

$$Q_U = M_e \, \overline{V}_e$$
$$q_u = m_e \, \overline{v}_e.$$

In other words, the parameters Q and q, seen from the domain U as charges, are really momenta in the domain e. We then have

$$F = \pm \, C \, \frac{Q_u \, q_u}{d^2} = \pm \, C \, \frac{M_e \, \overline{V}_e \, m_e \, \overline{v}_e}{d^2}. \tag{1}$$

This coincides with what we had considered for Newton's expression, and accordingly the main obstacle to unifying fields would have been overcome. In Figure 16, we saw the interpretation of phenomena in the light of these changes.

47

In summary:

(Newton)
$$F = \pm\, G\, \frac{M_u \overline{V}_u\, m_u \overline{v}_u}{d^2}$$

(Coulomb)
$$F = \pm\, C\, \frac{M_e \overline{V}_e\, m_e \overline{v}_e}{d^2} \quad ;$$

and in general:

$$F = \pm\, K\, \frac{M_i \overline{V}_i m_i \overline{v}_i}{d^2} \quad . \tag{2}$$

$$G = 6.67 \times 10^{-11}\ \text{New. m}^2/\text{kg}^2$$
$$C = 9 \times 10^9\ \text{New. m}^2/\text{coul}^2$$

In order to use the gravitation constant in Coulomb's equation, we have to touch upon other points. We will return to this at the end of the chapter.

Let us look at what is understood as "unifying fields." It would imply having a theory that would not only account for the effects of gravitational and electromagnetic fields but also the quantification or discrete nature of the electric charge, the atomicity of matter, and phenomena for which the corresponding equations incorporate Planck's constant. Einstein's work in his later years was already pointing the way to this ambitious goal. During the years 1945 and 1946 he tried to generalize his general theory of relativity by considering the g_{ik} components of the fundamental tensor as complex, with the real part being symmetrical while the imaginary one would have an asymmetric nature. This theory, however, would apply only to fields having a center of symmetry. In 1950, Einstein considered the fundamental tensor g_{ik} with real components as not being symmetrical—that is, that g_{ik} would be different from g_{ki}.

In this way differential equations with partial derivatives may be obtained, which are numerically sufficient, in principle, to solve any problem. The structure, however, of these equations is so complicated that, to date, it has not been possible to extract a single consequence from the theory to compare with experience. Finally Einstein, at the age of seventy-five in 1954, announced a new unified field theory, but once again, the equations are so complex that solving them is an almost superhuman task, making the extraction of verifiable consequences from the theory virtually impossible.

We have pointed out the above in view of Einstein's reputation and to make mention of his efforts to unify the fields. After a life devoted to science and notable for the success he achieved, Einstein probably felt himself well placed to culminate his efforts by leaving a series of general equations that would serve validly for all physics. He was never able to achieve this goal; we do know that he was convinced that a deeper knowledge of mathematics would, one way or another, offer access to an acceptable formulation of the matter, although his legacy was, in fact, a most complicated one.

We can reasonably conclude that unifying any set of things at all must demand an intimate knowledge of those things, and anything less than this is no more than speculation and hypothesis. I think that the interpretation or approach we have employed for gravitational and electrical phenomena serves to clarify the array of problems we face in our endeavors to unify the fields. Even so, it is also my opinion that we will never manage the total unification, insofar as the range of our knowledge of what goes on in certain domains—mainly the UU and ee domains—will for years and probably centuries of scientific progress inevitably fall short; as it goes forward there will be a constant need for modifying and amending the unification formulae obtained.

Consequently, if we take the expression (2), given above, as true, we can try to explain electromagnetic phenomena—in particular, that of light—later in Chapter IV.

We shall now try to show how relativity is applicable up to now in the domain U. This concerns relativity in events on earth and in the universe, where an adequate description requires us not to consider time as absolute. In other words, two events that may appear as simultaneous are not, in fact, so because the information for determination of the moments of their occurrence is brought to us by means of light. This is the fastest-traveling emanation in the universe, and its velocity is taken as a constant. Since the velocity of light is not infinite but simply very high, events that appear as simultaneous cannot be said to be so, and there is no other way of obtaining more reliable data for the time determination. The only solution is to set up correction formulae for all physical phenomena. In this way, relativistic physics was used to revize, clean up, and generally to arrive at more precise formulations of the formulae of classical mechanics. The relativity of simultaneity also gives rise to a relativity in distance measurements of segments located along the direction of the relative velocity.

In order to make such calculations, it is necessary to use inertial systems with uniform rectilinear motion along with transformation equations.

In this kind of dynamics, it is usual to use the term *impulse* for up to now what we have called *momentum*. The two following conditions

49

must also be satisfied: (1) The relativistic dynamics equations must conserve their form when passing from one system to another by application of the Lorentz transformation equations; this is known as the covariance of natural laws. (2) The new equations must coincide with the Newtonian equivalents for velocities, which are small compared with that of light.

This theory may be applied to any phenomena in the domain U—or, rather, phenomena observed from U—and would include, for instance, the energy transformation of a beam of light, the pressure of light, uniformly accelerated motion, hyperbolic motion, electromagnetic fields, the behavior of an electric charge moving in a field, entropy invariance, *et cetera*.

There are, in fact, proofs of this restricted relativity theory, such as Fizeau's experiment, the radiation of mass with velocity in radioactive phenomena, or in the case of artificially accelerated electrons (which have a unit mass value of 1.494×10^{-3} ergs)—all, let us remember, observed from U.

Later on, with the general theory of relativity, it was possible to establish that a small region Q of a gravitational field where bodies fall with an acceleration of g is equivalent to a system K' moving with a uniformly accelerated motion and acceleration g, with respect to an inertial system K.

Having a knowledge of the non-Euclidean geometries and their validity—mainly for the vast sidereal distances—we can apply these to study the gravitational field and arrive at a geometry for such a field.

Thus a knowledge of both spatial and time coordinates has enabled us to set up a spatial physics in the gravitational fields in U.

The main experimental evidence supporting this approach has been that of the shift in Mercury's perihelion, the bending of light rays in the neighborhood of the Sun, and the red shift in the spectra of light from the Sun and other stars.

RELATIONSHIPS BETWEEN CONSTANTS

The effects of gravitational and electromagnetic fields and the discrete or quantum nature of the electric charge are interpreted in Chapter II.

We shall now pass to phenomena in the description of which Planck's constant appears. We should, however, first point out that (1) the principle of the conservation of energy and that of momentum is the same thing, and we call to mind the principle of conservation of mass, and (2) recalling the value v/ν from the table in Chapter I—*viz*, unity—we can take it into account in light of the previous point, since if $V_u/V_e = 1$ $\Rightarrow Cm_u/Cm_e = M_u/m_e$, hence masses are energy carriers, remaining unchanged in passing from one domain to another. Here again we see that the energies represented by momenta are proportional to the mass, and at constant mass, the greater the mass the greater the energy for the same momentum per unit mass.

Therefore v/ν must be equal to 1, and since v = 0.6 x 10^7 cm/sec, then ν must be the same value—0.6 x 10^7 cm/sec—for the principle of the conservation of momentum to remain unviolated.

With the velocity ν equal to 0.6 x 10^7 cm/sec and Γ equal to 3.77 x 10^{-4} sec, the distance of this velocity is as follows:

$$\nu = \frac{e}{\Gamma}$$

e = ν Γ = (0.6 x 10^7) (3.77 x 10^{-4}) = 2.2 x 10^3 cm.

This value corresponds to the wavelength, in the electromagnetic spectrum, of short-wave radio transmissions.

What follows is an effort to give a physical sense to the universal gravity constant in the light of the new criteria and to relate it with h, E, and m, since these values are significant.

We shall use the following deductive sequence:

(1) We consider that the constant of universal gravity G is equal to 6.67 x 10^{-8} dyne cm^2/gr^2, hence 6.67 x 10^{-11} Nm^2/kg^2 is the gravity constant in the domain U resulting from the gravity in domain e observed from U.

(2) We consider K (equal to 8.84 x 10^2 gr^2 cm/sec) the gravity constant in e resulting from the gravity in ee, observed from e.

(3) We consider the product, KG, equal to G' as the gravity constant of ee, observed from U.

(4) If we recall that the C^2 value from the expression E = mc^2 gives a good result and that, at present, there is no reason to alter it (it has been justified though not placed on an entirely solid basis), we shall endeavor to find the expression for E as a function of the domain passage concept.

(5) If we also take Planck's constant as the minimum action that an element of the domain e must have to pass to the domain U and give rise to phenomena discernible in U.

(6) An example would be how to find the capacity of action of an electron.

Let us calculate its momentum Cm.

Cm = m$_e$v$_e$ = 9.18 x 10^{-28} gr x 0.6 x 10^7 cm/sec = 5.4 x 10^{-21} gr cm/sec

and $\dfrac{Cm}{h}$ = $\dfrac{5.4 \text{ x } 10^{-21} \text{ gr cm/sec}}{6.6 \text{ x } 10^{-26} \text{ gr cm}^2/\text{sec}}$ = 0.8 x 10^5 cm^{-1} then

Cm$_e$ = 0.8 x 10^5 cm^{-1} h.

(7) If G is the e to U domain passage constant and h is the minimum amount of momentum to pass from e to U, then G/h gives us the unit e to U domain passage constant from the product of mass by total velocity.

$$\frac{G}{h} = \frac{6.7 \times 10^{-8} \text{ dyne cm}^2/\text{gr}^2}{6.6 \times 10^{-34} \text{ Joule/sec}} = \frac{10^{-8}}{10^{-34} \times 10^7} = 10^{19} \text{ cm sec/gr}^2$$

If we assume that G/h is the action unit for each dm_{ei}

$$m_u = \sum_{i=1}^{i=n} dm_{ei}$$

$$E_u = \sum_{i=1}^{i=n} dE_{ei}$$

Thus, we can write: $E_U = m_e\, G/h$, which is

$$E = Km\, \frac{G}{h}$$

(8) We have from Chapter II that the change of scale in passing from U to e (for distances and times) is of the order of C^2 or $(KG)/h$.

(9) We can proceed then to say $E = mc^2 = (Km\, G)/h$. We use K because of the light (and since we are using its velocity—*i.e.*, it is from the domain ee) and G because we measure the phenomenon, and thus arrive at U.

Also, we have

$$C = \sqrt{\frac{KG}{h}}$$

$$\frac{Eh}{m} = 58.96 \times 10^{-6} \text{ gr cm}^4/\text{sec}^3 = \text{constant}$$

$KGm = Eh$ (Energy x constant = mass x constant)

thus $\boxed{G'm = Eh}$.

This expression would be applicable for e observed from U; and further:

$$\frac{m}{h} = \frac{E}{G} .$$

In the above we see more sharply the equivalence between mass and energy where we understand mass as the energy carrier through $\overline{V}/\overline{v}$. In the last expression in particular, it would appear that the relationship means the smallest amount of mass carrying the smallest amount of energy necessary to exit from e and make itself known by means of measurable phenomena in U. In the same expression m can be replaced by E (where $E = f (m, \overline{V}/\overline{v})$ in the case of each particular phenomenum.

I realize that to deduce the values of G and K and finally G' must be very difficult insofar as it concerns the resultant of all the gravitational fields, and, therefore, the complexity involved must be mind boggling. Accordingly, it is more convenient to work with the G value obtained by experiment and K as deduced from the corresponding equations.

If the G value originates in the e level and the K value in the level ee, it is reasonable to suppose that there must be some value originating in the level U and which can be observed in the level UU.

In order to appreciate the complexity of the problem, we would have to find (a) the resultant of the Sun's planetary system, already complex, with reference to its energy contribution in the universe—or, in other words, to its participation in the overall energy condition in U; this process would have to be carried out for the entire domain in U. (b) If we were to arrive at a value, theoretically impossible to find with precision, such a value would have to be a dimensional associated passage expression, which then would indeed be the gravitational constant of U, observed from UU.

The nature of this problem is such that it prevents us trying anything similar for the atomic domain since we are not sufficiently familiar with phenomena at the e and ee levels. This is to say that I believe the analytical deduction of the constants G and K is beyond the possibilities of science as it stands today. The best we could hope for would be to interpret their approximate meaning, and if this fits, then we would be on the right track.

This is all I can say in support of the acceptance of K and that, if G were accepted by means of experimental determination, the same could be done for K in tests that could be carried out, apart from the different expressions that incorporate this constant.

In this way we have a relationship between the constant of universal gravity—with some meaning—Planck's constant and the speed of light (used as a constant in U) in a single expression.

If physics were unified, we would have expressions for any phenomena as a function of the parameters; such phenomena today are still classified in the different categories of physics. In other words, with the unification of physics into a single whole, we would have a single complete and closed set of mutually compatible expressions.

There is a lot of work involved in adaptation, in locating exactly where the various phenomena actually take place—in the cosmos, the microcosmos, and so forth—and also in the interpretation of the real meaning of the different constants. This will take time to comprehend and to become accustomed to thinking of phenomena in the unified field.

There is no doubt that these processes of location and interpretation will bring definite progress in the knowledge and understanding of the reality of phenomena.

The complexity of the unified field formulae compared with the more simple expressions we use at present should not mean that these formulae ought to be abandoned. On the contrary, the simplest expressions are always those sought for solving problems.

In line with the following list of the meaning of symbols we set up Figure 27, having the usual expressions and the unified field expressions. Thereafter we shall develop ten examples, although a discussion of the significance of the resulting expressions falls beyond the scope of this essay.

UNIFIED FIELD EXPRESSIONS

λ = wave length
c = speed of light
ν = frequency
h = Planck's constant
m = mass
v = velocity
E = energy
G = gravity constant of domain U
K = gravity constant of domain e
G' = total gravity constant
g = acceleration of gravity
d = radius of the earth
M = mass of the earth
μ = electrical permeability
ϵ = electrical induction
C = specific heat
Q = heat capacity

54

q = electric charge
V = voltage
F = force
p = pressure
ρ = density
v′ = volume
L = length
$\bar{a}$ = characteristic acceleration
P = weight

	Usual expression	Unified field expression
λ	$\dfrac{h}{m.v}$	$\dfrac{G'}{c^2.m.v}$, $\sqrt{\dfrac{K.G}{m.v^2.\nu}}$, $\dfrac{KG}{c^2.m.v}$, etc.
ν	$\dfrac{E}{h}$	$\dfrac{G'.m}{h^2}$, $\dfrac{c^4.m}{KG}$, $\dfrac{K.G.m}{h^2}$, etc.
h	$\lambda.m.v$ $\dfrac{E}{\nu}$	$\dfrac{K.G}{c^2}$, $\dfrac{g.K.d^2}{c^2.M}$, $\sqrt{\dfrac{G'.m}{\nu}}$, $\dfrac{G'}{c^2}$, $\sqrt{\dfrac{K.G.M}{\nu}}$, $\dfrac{G'.\mu_0.\epsilon_0}{E.\lambda.v}$, $\mu_0\,\epsilon_0\,KG$, $\mu_0.c_0.G'$, etc.
E	$m.c^2$	$\dfrac{m.G'}{h}$, $m\dfrac{K.G}{h}$, $\dfrac{G'.\mu_0.\epsilon_0}{\lambda.v.h}$, etc.
c	$\sqrt{\dfrac{m}{E}}$ $\sqrt{\dfrac{1}{\mu_0.\epsilon_0}}$	$\sqrt{\dfrac{G'}{h}}$, $\sqrt{\dfrac{KG}{h}}$, $\left(\dfrac{K.G.\gamma}{m}\right)^{-\frac{1}{4}}$, $\sqrt{\dfrac{g.k.d^2}{h.M}}$, $\dfrac{E.v^2}{2q.V}$, etc.
G′	$K.G$	$h.c^2$, $\dfrac{h}{\mu_0.\epsilon_0}$, $\sqrt{\dfrac{E.\lambda.v.h}{\mu_0.\epsilon_0}}$, $\dfrac{E.h.v^2}{2.q.V}$, etc.
G	$\dfrac{F.d^2}{M.m}$	$\dfrac{h.c^2}{K}$, $\dfrac{h}{K.\mu_0.\epsilon_0}$, etc.
g	$G\,\dfrac{M}{d^2}$	$\dfrac{h.c^2.M}{K.d^2}$, etc.

Figure 27

55

	Usual expression	Unified field expression
K	$\dfrac{G'}{G}$	$\dfrac{h}{G.\mu_0.\epsilon_0}$, $\dfrac{h.c^2}{G}$, $\dfrac{\nu^2.h^3}{m^2.G}$, etc.
m	$\dfrac{E}{c^2}$	$\dfrac{G'.\mu_0.\epsilon_0}{\lambda.v}$, $\dfrac{E.h}{G'}$, etc.
$\dfrac{p}{\rho.g}$	$\dfrac{d^2}{GM}$	$\dfrac{m.K.d^2}{E.h.M}$, etc.
$p.v'$	constant	$\dfrac{E.h.\bar{a}}{G'}$ = constant, etc.
C	$\dfrac{Q}{m.\Delta t}$	$\dfrac{Q.G'}{E.h.\Delta t}$, etc.
$q.V$	$\dfrac{1}{2}.m.v^2$	$\dfrac{E.K.G.v^2}{2.h.c^4}$, etc.
F	$G\dfrac{M.m}{d^2}$	$\pm\, G\,\dfrac{M.m.\overline{V}}{d^2.\bar{v}}$, $\pm\left(\dfrac{h^2.E_1E_2.\overline{V}}{K^2.G.d^2.\bar{v}}\right)$, etc.
F	$\pm\, C\dfrac{Q.q}{d^2}$	$\pm\, C\dfrac{M.m.\overline{V}}{d^2.\bar{v}}$, $\pm\,\dfrac{C}{G}\left(\dfrac{h^2.E_1E_2\overline{V}}{K^2.G.d^2.\bar{v}}\right)$, etc.
p	m.g	$\dfrac{E.h^2.c^2.M}{G.K^2.d^2}$, $\nu\,\dfrac{M}{c.K.d^2}$, etc.

Figure 27 (continued)

At low energy, we have:

$$F = \pm\, G\,\frac{Mm\,\overline{V}/\bar{v}}{d^2} = \pm\, G\,\frac{\dfrac{E_1E_2\,h^2}{G^2\,K^2}\overline{V}/\bar{v}}{d^2} = \pm\,\frac{G}{c^4}\,\frac{E_1E_2\,\overline{V}/\bar{v}}{d^2}$$

$$= \pm\, G\,\frac{\dfrac{E_1E_2\,h^2}{G^2\,K^2}\overline{V}/\bar{v}}{d^2} = \pm\,\frac{E_1E_2\,h^2\,\overline{V}}{G\,K^2\,d^2\,\bar{v}}$$

$$= \pm\,\frac{h^2}{G'K}\,\frac{E_1E_2\,\overline{V}/\bar{v}}{d^2} = \pm\,\frac{1}{G}\,\frac{h^2}{K^2}\,\frac{E_1E_2\,\overline{V}}{d^2\,\bar{v}}\;;$$

and at high energy:

$$F^* = \pm\, C\, \frac{Mm\, \overline{V}/\overline{v}}{d^2} = \pm\, \frac{C}{c^4}\, \frac{E_1 E_2\, \overline{V}/\overline{v}}{d^2} = \pm\, C\, \frac{\dfrac{E_1 E_2\, h^2}{G^2 K^2}\, \overline{V}/\overline{v}}{d^2}$$

$$= \pm\, \frac{c.h^2}{G'^2}\, \frac{E_1 E_2\, \overline{V}/\overline{v}}{d^2} = \pm\, C\, \frac{E_1 E_2\, h^2\, \overline{V}}{G^2\, K^2\, d^2\, \overline{v}}$$

$$= \pm\, \frac{C.}{G^2} \cdot \frac{h^2}{K^2} \cdot \frac{E_1 E_2\, \overline{V}}{d^2\, \overline{v}}\, .$$

$$\boxed{F^* = F\, C/G}$$

For the same phenomenon the last expressions should be equal, in which case, significantly, we finally obtain the equality of the gravitational and electrical constants:

$$\frac{1}{G} = \frac{C}{G^2}\text{, whence } G = C.$$

Now, this expression explains the phenomenon but does not quantify it. For such cases, the corresponding equation must be employed, and the difference between them indicates the difference between the gravitational and electrical energy levels.

The G value: $G\colon 58.96 \times 10^{-6}\ cm^4\ gr/sec^3 = KG$

The C value: $c\colon \sqrt{\dfrac{G'}{h}} = \sqrt{\dfrac{58.96 \times 10^{-6}\ cm^4\ gr/sec^3}{6.6 \times 10^{-26}\ cm^2\ gr/sec}}$

$$= 2.9 \times 10^{10}\ cm/sec$$

The E value: $E\colon m\, \dfrac{G'}{h} = \dfrac{1\ gr \times 58.96 \times 10^{-6}\ cm^4\ gr/sec^3}{6.6 \times 10^{-26}\ cm^2\ gr/sec}$

$$= 8.93 \times 10^{20}\ joule/sec$$

Calculation of g:

$$F = mg$$
$$g = F/m$$
$$F = \pm\, \frac{h\, c^2\, Mm\, \overline{V}}{K\, d^2\, \overline{v}}\, 1\,;\ \text{with}\,\frac{\overline{V}}{\overline{v}} = 1\ \text{and}\ \frac{h}{K} = \frac{G}{c^2}$$

57

$$g = \frac{h\,c^2\,Mm}{K\,d^2\,m} = \frac{G\,c^2\,M}{c^2\,d^2} = G\,\frac{M}{d^2}$$

$$g = 6.7 \times 10^{-8} \text{ dyne cm}^2/\text{gr}^2 \times \frac{5.98 \times 10^{27} \text{ gr}}{40.7 \times 10^{16} \text{ cm}^2}$$

$$= 9.8 \text{ m/sec.}^2$$

And then we can write:

$$g = \frac{h\,c^2\,M}{K\,d^2}$$

$$c = \sqrt{\frac{g\,K\,d^2}{h\,M}} \qquad\qquad h = \frac{g\,K\,d^2}{c^2\,M}\;.$$

The gravitational and electromagnetic fields are connected by the different parameters. To check this:

$$g = \frac{h^2\,c^2\,M}{K\,d^2}$$

$$\frac{M}{d^2} = \frac{5.98 \times 10^{27} \text{ gr}}{40.7 \times 10^{16} \text{ cm}^2} = 1.469 \times 10^{10} \text{ gr/cm}^2$$

$$\frac{h\,c^2}{K} = \frac{6.6 \times 10^{-26} \text{ gr cm}^2/\text{sec} \times 9 \times 10^{20} \text{ cm}^2/\text{sec}^2}{8.84 \times 10^2 \text{ gr}^2 \text{ cm/sec}}$$

$$= \frac{59.4 \times 10^{-6} \text{ gr cm}^4/\text{sec}^3}{8.84 \times 10^2 \text{ gr}^2 \text{ cm/sec}} = 6.71 \times 10^{-8} \frac{\text{cm}^3}{\text{sec}^2 \text{ gr}}$$

$$g = \frac{h\,c^2}{K}\,\frac{M}{d^2} =$$

$$= 1.469 \times 10^{10} \text{ gr/cm}^2 \times 6.71 \times 10^{-8} \text{ cm}^3/\text{sec}^2 \text{ gr}$$

$$= 9.8 \frac{\text{m}}{\text{sec}^2}\;.$$

All the other unified field equations should be verified in the same way so as to ensure they have been set up in adequate fashion. The dimensions of K are those of a momentum with the mass taken twice: this is specially notable if we take into account that this is the constant that jumps two domains.

We note that $g = (hE)/(Kd^2)$; this means that an initial interpretation could be that g is a function of the total energy (E) of the system; its field decreases as the inverse square (d^2) of the distance; it is proportional to h and inversely proportional to K—*i.e.*, to two constants that originate at the atomic level.

Bernoulli's theorem,

$$\frac{p_i}{\rho g} + y_1 = \frac{p_2}{\rho g} + y_2$$

$$p = \frac{F}{L^2} = \frac{\dfrac{h^2 \, c^2 \, Mm \, \overline{V}}{K \, d^2 \, \overline{v}}}{L^2}$$

$$= \frac{h \, c^2 \, Mm}{K \, d^2 \, L^2}_1 ; \text{ with } \frac{\overline{V}}{\overline{v}} = 1$$

$$g = \frac{h \, c^2 \, M}{K \, d^2}$$

$$\rho = \frac{P}{L^3} = \frac{mg}{L^3} = \frac{\dfrac{E \, h}{G'} \cdot \dfrac{h^2 \, c^2 \, M}{K \, d^2}}{L^3}$$

$$= \frac{\dfrac{E \, h}{h \, c^2} \cdot \dfrac{h \, c^2 \, M}{K \, d^2}}{L^3} = \frac{\dfrac{E \, h \, M}{K \, d^2}}{L^3}$$

$$\frac{p}{\rho g} = \frac{\dfrac{h \, c^2 \, Mm}{K \, d^2 \, L^2}}{\dfrac{\dfrac{E \, h \, M}{K \, d^2}}{L^3} \cdot \dfrac{h \, c^2 \, M}{K \, d^2}}$$

$$= \frac{h \, c^2 \, Mm}{K \, d^2 \, L^2} \cdot \frac{L^3}{\dfrac{E \, h \, M}{K \, d^2}} \quad \frac{K \, d^2}{h \, c^2 \, M}$$

$$= \frac{h \, c^2 \, M \, m \, L \, K \, d^2 \, K \, d^2}{k \, d^2 \, E \, h \, M \, h \, c^2 \, M} = \frac{m \, K \, d^2 \, L}{E \, h \, M}$$

$$= \frac{m \, K \, d^2 \, L}{\dfrac{m \, G' \, h \, M}{h}} = \frac{K \, d^2 \, L}{G' \, M} = \frac{d^2 \, L}{G \, M} \quad ,$$

can be compared with

$$\frac{p}{\rho g} = \frac{F/L^2}{p/L^3 \, G \, \dfrac{M}{d^2}} = \frac{F}{L^2} \, \frac{L^3}{p} \, \frac{d^2}{GM} = \frac{m \, a}{L^2} \, \frac{L^3}{mg} \, \frac{d^2}{GM} = \frac{d^2 \, L}{G.M} \, .$$

59

Then, we can write

$$\frac{m_1 \, K \, d^2 \, L}{E \, h \, M} + y_1 = \frac{m_2 \, K \, d^2 \, L}{E \, h \, M} + y_2 \quad .$$

Using Boyle's law, F represents the internal pressure in a vessel.

$$F = m \, \bar{a}$$

$$p \, v' = \frac{F}{L^2} \, L^3 = m \, \bar{a} \, L = \text{constant}$$

$$p \, v' = \frac{E \, h \, \bar{a}}{G'} \, L = \frac{m \, c^2 \, h \, \bar{a} \, L}{h \, c^2}$$

$$= m \, \bar{a} \, L = \text{constant}$$

And we can write

$$p \, v' = \frac{E \, h \, \bar{a} \, L}{G'} = \frac{E \, h \, \bar{a} \, L}{K \, G} = \text{constant} \quad .$$

Regarding electromagnetic waves, we know that, in Coulomb's law, $K = 9 \times 10^{19}$ New m^2/coul^2 and that for the specific induction capacity of the medium,

$$\epsilon_o = \frac{1}{4 \, \pi \, K} = 8.85 \times 10^{-12} \text{ coul}^2/\text{New m}^2 \quad .$$

Now, with the permeability μ, the greatest success of the electromagnetic theory was to determine that light has electromagnetic characteristics, having a speed C, given by

$$c = \sqrt{\frac{1}{\mu \, \epsilon}} \quad ,$$

and for light in a vacuum,

$$c = \sqrt{\frac{1}{\mu_o \, \epsilon_o}} = 3 \times 10^{10} \text{ cm/sec.}$$

We now write

$$c = \sqrt{\frac{G'}{h}} = \sqrt{\frac{1}{\mu_o \, \epsilon_o}} \quad ;$$

then

$$\frac{h}{G'} = \mu_o \, \epsilon_o$$

$$\frac{h}{h \, c^2} = \mu_o \, \epsilon_o$$

$$c^2 = \frac{1}{\mu_o \, \epsilon_o}$$

$$c = \sqrt{\frac{1}{\mu_o \, \epsilon_o}} \, .$$

In view of the similarity we can now write

$$\frac{h}{K \, G} = \mu_o \, \epsilon_o \quad \text{and} \quad K = \frac{h}{G \, \mu_o \, \epsilon_o} \quad ;$$

$$G = \frac{h}{K \, \mu_o \, \epsilon_o} \quad \text{and} \quad G' = \frac{h}{\mu_o \, \epsilon_o}$$

$$h = \mu_o \, \epsilon_o \, K G' = \mu_o \, \epsilon_o \, G^1$$

For example:

$$h = \mu_o \, \epsilon_o \, G' =$$

$$1.26 \times 10^{-7} \times 8.85 \times 10^{-15} \times 5.9 \times 10^{-5} \, \frac{cm}{gr} \times gr \, \frac{sec^2}{cm^3} \times gr \, \frac{cm^4}{sec^3}$$

$$= 65.79 \times 10^{-27} \, gr \, cm^2 / sec = 6.6 \times 10^{-26} \, gr \, \frac{cm^2}{sec}$$

$$G = \frac{h}{K \, \mu_o \, \epsilon_o}$$

$$= \frac{6.6 \times 10^{-26} \, gr \, cm^2 / sec}{8.84 \times 10^2 \, gr^2 \, \dfrac{cm}{sec} \times 1.26 \times 10^{-7} \, \dfrac{cm}{gr} \times 8.85 \times 10^{-15} \, gr \, \dfrac{sec^2}{cm^3}}$$

$$= \frac{6.6 \times 10^{-26} \, gr \, cm^2 / sec}{99 \times 10^2 \times 10^{-7} \times 10^{-15} \, gr^2 \, \dfrac{cm}{sec} \, \dfrac{cm}{gr} \, gr \, \dfrac{sec^2}{cm^3}}$$

$$= \frac{6.6 \times 10^{-26} \text{ gr cm}^2/\text{sec.}}{99 \times 10^{-20} \text{ gr}^2 \ \dfrac{\text{sec}}{\text{cm}}}$$

$$= \frac{6.6 \times 10^{-26} \text{ gr cm}^2/\text{sec}}{0.99 \times 10^{-18} \text{ gr}^2 \text{ sec/cm}} = 6.67 \times 10^{-26} \times 10^{18} \text{ gr} \ \frac{\text{cm}^2}{\text{sec}} \ \frac{\text{cm}}{\text{sec gr}^2}$$

$$= 6.67 \times 10^{-8} \text{ gr} \ \frac{\text{cm}}{\text{sec}^2} \ \frac{\text{cm}^2}{\text{gr}^2} = 6.67 \times 10^{-8} \text{ dyne cm}^2/\text{gr}^2.$$

And also

$$m = \frac{G' \ \mu_o \ \epsilon_o}{\lambda \ v} = \frac{E \ h}{G'}$$

$$E = \frac{G'^2 \ \mu_o \ \epsilon_o}{\lambda \ v \ h}$$

$$h = \frac{G'^2 \ \mu_o \ \epsilon_o}{E \ \lambda \ v}$$

$$G' = \sqrt{\frac{E \ \lambda \ v \ h}{\mu_o \ \epsilon_o}} \ .$$

When measuring specific heat,

$$C = \frac{\dfrac{Q}{\Delta t}}{m} = \frac{Q}{m \ \Delta t}$$

$$m = \frac{E \ h}{G'}$$

$$G' = KG$$

$$\boxed{C = \frac{Q}{\dfrac{E \ h \ \Delta t}{G'}} = \frac{Q \ G'}{E \ h \ \Delta t} = \frac{Q \ K \ G}{E \ h \ \Delta t}} \ .$$

In the cyclotron, we have

$$\tfrac{1}{2} \ m.v^2 = q \ V$$

$$E = m. \frac{G'}{h} \qquad\qquad m = \frac{E \ h}{G'}$$

$$q\,V = \frac{E\,h\,v^2}{2\,G'}$$

$$G' = \frac{E\,\dfrac{KG}{c^2}\,v^2}{2\,q\,V} = \frac{E\,K\,G\,v^2}{2\,c^2\,q\,V}$$

$$h\,c^2 = \frac{E\,K\,G\,v^2}{2\,c^2\,q\,V}$$

$$h\,c^4 = \frac{E\,K\,G\,v^2}{2\,q\,V}$$

$$\boxed{q\,V = \frac{E\,K\,G\,v^2}{2\,h\,c^4}}$$.

And

$$G' = \frac{E\,h\,v^2}{2\,q\,V}$$

$$h\,c^2 = \frac{E\,h\,v^2}{2\,q\,V}$$

$$c = \sqrt{\frac{E\,v^2}{2\,q\,V}}$$.

We can also write the two voltage and current equations reduced to two wave equations in V and i:

$$\frac{\partial^2 V}{\partial t^2} = \frac{G'}{h}\,\frac{\partial^2 V}{\partial Z^2}$$

$$\frac{\partial^2 i}{\partial t^2} = \frac{G'}{h}\,\frac{\partial^2 i}{\partial Z^2}$$

Here we have voltage and electric current as a function of Planck's constant and that of total universal gravity. In this same fashion, on the basis of the unifying of the fields, the whole of physics can be revised. The standard kilogram may also be expressed as a function of a specified wavelength.

The following relativistic expressions in the unified field allow for the translation of information obtained in a reference inertial system into information valid for another system. Subsequently, equivalent inertial systems are considered on the basis of the supposition that the velocity of light is equal in all of them.

$$x' = \frac{x - v\,t}{\sqrt{1 - \dfrac{v^2}{c^2}}}$$

$$t' = \frac{t - \left(\dfrac{v}{c^2}\right) x}{\sqrt{1 - \left(\dfrac{v^2}{c^2}\right)}} \quad,$$

which we may write as

$$c^2 = \frac{G'}{h} = \frac{g\,K\,d^2}{h\,M} \text{ and } h = \mu_o\,\epsilon_o\,KG.$$

$$x' = \frac{x - v\,t}{\sqrt{1 - \dfrac{v^2\,h}{G'}}} = \frac{x - v\,t}{\sqrt{1 - \dfrac{v^2\,h\,M}{g\,K\,d^2}}} = \frac{x - t}{\sqrt{1 - v^2\,\mu_o\,\epsilon_o}} \, et \, cetera.$$

$$t' = \frac{t - \left(\dfrac{v\,h}{G'}\right) x}{\sqrt{1 - \dfrac{v^2\,h}{G'}}} = \frac{t - \left(\dfrac{v\,h}{G'}\right) x}{\sqrt{1 - v^2\,\mu_o\,\epsilon_o}} = \frac{t - (V\,\mu_o\,\epsilon_o)\,x}{\sqrt{1 - \dfrac{v^2\,h\,M}{g\,K\,d^2}}}$$

$$= \frac{t - (v\,\mu_o\,\epsilon_o)\,x}{\sqrt{1 - \dfrac{v^2\,\mu_o\,\epsilon_o\,G\,M}{g\,d^2}}} \, et \, cetera.$$

We should note that in the last expressions of x' and t' neither K nor G' appear.

Now, if x = 10 cm, v = 6 cm/sec, and t = 2 sec:

$$x' = \frac{x - vt}{\sqrt{1 - \dfrac{v^2}{c^2}}} = \frac{-2}{\sqrt{1 - \dfrac{36}{9 \times 10^{20}}}} = \frac{-2}{\sqrt{1 - \left(\dfrac{36}{9} \times 10^{-20}\right)}}$$

$$= \frac{-2}{\sqrt{1 - (4 \times 10^{-20})}} \ .$$

Let's check this:

$$x' = \frac{10 - 6 \times 2}{\sqrt{1 - \dfrac{36 \, h}{G'}}} = \frac{10 - 12}{\sqrt{1 - \dfrac{36 \times 6.6 \times 10^{-26}}{58.96 \times 10^{-6}}}} = \frac{-2}{\sqrt{1 - \dfrac{238 \times 10^{-26}}{58.96 \times 10^{-6}}}}$$

$$= \frac{-2}{\sqrt{1 - (4 \times 10^{-26} \times 10^6)}} = \frac{-2}{\sqrt{1 - (4 \times 10^{-20})}} \ .$$

It may also be said that without G' substitutions may be made; for example:

$$E_c = \frac{1}{2} m v^2 = \frac{1}{2} \frac{E}{c^2} v^2$$

$$= \frac{E v^2}{2 \sqrt{\dfrac{1}{\mu_o \, \epsilon_o}}}$$

But let us now look at what happens in the case of quantum mechanics and relativity. We have already seen that in the parallel worlds of U and e, gravitational phenomena are similar and therefore may, in their respective domains, be described with relativistic equations; to do this, we had to correct the universal gravity constant. If we are to take the domain e into account, then consequently the unification of the fields and finally the study with relativity and quantum mechanics are compatible. There is still, however, the problem of an adequate determinism in quantum mechanics for describing phenomena in the domain e in the same way we are accustomed to doing so in the domain U, but if we direct ourselves by the change of time scales, we realize how difficult it is. The substitutions in the equations, however, serve for a better understanding of these phenomena since we have more interchangeable formulae with K than without it.

With G′ a complete compatible system of equations for all physics can be formed. This cannot be done without G′.

It can be shown that it is a closed, complete, and compatible system by using the following routes:

(1) dimensional calculus;

(2) search for a case that does not fit;

(3) by a deductive progression passing through all areas of physics, thus verifying an uninterrupted line.

Bearing in mind the last criterion mentioned above, we are already aware that without G′ the most apparent discontinuity in the deductive progression was between the gravitational and electromagnetic fields. The solution to a problem such as this must bring, too, a compatibility between quantum and relativistic mechanics.

We shall now return to the matter of the constant in Coulomb's equation, not forgetting that it corresponds to a high-energy phenomenon and that the gravity constant corresponds to a relatively low-energy one. These two should be related if the values indicated a proportionality with respect to the energy state.

If we consider the quotient C/G as the number of times that G is contained in C, then in Coulomb we can use the universal gravitation constant multiplied by this quotient; hence, the new constant C′ will be $C' = G \, C/G$, or $C' = C$.

As for K, we have the problem that F is in dynes, Newtons, *et cetera*, and if we use K, F will come in gr x dynes or kg Newtons.

This shows that the use of K is out of place and further confirms that electric charge originates in the domain e rather than in e and ee, as is the case for gravity.

This being the case, GK is the total gravitational constant from ee to U; if we divide by the unitary constant of ee, we obtain the constant for use in U for the calculation of F, thus $GK/K = G$. If this, in fact, concerns the same phenomenon, then the following condition ought to be fulfilled; $G \, C/G = GK/K$—i.e., $C/G^2 = 1/G$; hence $G = C$. If we were to use G in Coulomb, the body would not be electrified and hence Coulomb would not be applicable. If we were to use C in Newton, the domain U would have the energy of a universe seen from UU, just as we detect and measure electricity in e.

Such a U domain, with energy equal to that of electrification in e, is almost impossible to calculate because we would need a constant C#, detectable from UU for the high-energy case in domain U. In order to determine C#, the corresponding F#, value is just what we would need to know to enable us to find $\overline{V}$ and $\overline{v}$ in U for such a case. The dimensions of C# may well be $(gr^3 cm^4)/(sec^3 Coul^2)$.

We cannot do such a calculation; so, for now, we take the example of the calculation of the electrostatic repulsion force between two alpha particles separated by a distance of 10^{-11} cm.

Each alpha particle has a double positive charge; thus we have e = 2 x 1.6 x 10^{-19} = 3.2 x 10^{-19} Coul.

$$-F = 9 \times 10^9 \; \frac{New \; m^2}{Coul^2} \; \frac{3.2 \times 10^{-19} \; Coul \times 3.2 \times 10^{-19} \; Coul}{(10^{-13} \; m)^2}$$
$$= 9.180 \; dynes$$

$$\boxed{\pm \, F = C \, \frac{M \, m \, \overline{V}}{d^2 \, \overline{v}}}$$

M.m = 8 x 16.4 x 10^{-25} gr^2 = 2.69 x 10^{-23} gr^2

$$\frac{\overline{V}}{\overline{v}} = \frac{10.24 \times 10^{-38} \; Coul^2}{2.69 \times 10^{-23} \; gr^2} = 3.8 \times 10^{-15} \; Coul^2/gr^2$$

M m $\overline{V}/\overline{v}$ = 10.24 x 10^{-38} $Coul^2$; then

$$-F = 9 \times 10^9 \; New \, m^2/Coul^2 \times \frac{2.69 \times 10^{-23} \; gr^2 \times 3.8 \times 10^{-15} \; Coul^2/gr^2}{(10^{-13} \; m.)^2}$$
$$= 9.180 \; dynes$$

We remember that the dimensions of MV.mv we $(gr^2 cm^2)/sec^2$ and that 1 $Coul^2$ = $(3.28 \times 10^{-2})^2$ $(gr^2 \; cm^3)/(sec^2)$, taking as a starting point the momentum of the electron and multiplying by the amount of electrons necessary to make up a Coulomb.

In this high-energy process, the sign is determined by consideration of the prior condition of the system. Conventionally we put it in F. The inversion, $\overline{V}/\overline{v} < 1$, is equivalent to repulsion, and this is due to its being an oversaturated planetary system in domain e.

Chapter IV

THE NATURE OF LIGHT

On the basis of what we saw in Chapter II we could locate the photon in the domain ee, as a minuscule mass with a rotating element, a quasi-tron, moving around it. The whole photon system, on moving the elements of the e domain, produced a perturbation of the potential fields. This perturbation, when measured by us from the U domain in terms of wavelength, constitutes the phenomenon we call light.

We will not consider the other phenomena to have electromagnetic characteristics so as not to complicate the presentation. The interpretation given here should, however, adequately apply to such other phenomena.

I would like to emphasize that if light is a phenomenon having the above-mentioned characteristics, then, in the absence of fields for it to perturb in its passing through, it should not be visible; in other words, light does not illuminate. If it were to illuminate in the virtual absence of matter—in space, for example—we would be unable to see the firmament. But it does illuminate, and this we observe through the atmosphere.

It is a good exercise to consider a simple light bulb, starting from the generation taking place in the filament heated by an electric current, to show that, in the absence of perturbable potential fields, the filament would be invisible. For this we must discard the potential fields of the eye itself, measuring apparatus, and those of that part of the filament which generates no light—i.e., a part of the filament perturbed by light generated in another section of it.

It is a little complex; were it simple, this concept would have been put to good use long ago. In fact, it arises from taking the origin of light as the ee domain.

On this basis, if light is the discharge of a planetary unit from the ee domain, it must have an initial or starting energy. Now, as far as change of a particular sign it may have had in the domains is concerned, light is neutral and its own potential would be of gravitationally saturated characteristics, which fact allows it to move through potential fields without a significant loss of energy. It would also explain the constancy of its velocity, but it makes us consider that its velocity, of necessity, must decrease when it moves freely through space or passes through bodies that are transparent to it.

The velocity of light must also abruptly drop to zero upon its being absorbed by some body, and if it is absorbed, its velocity decreases at a particular moment; if it decreases, it is not constant. In other words, the constancy of the velocity of light is valid, measured from the domain U for astronomical phenomena and as a constant in formulae; as an intrinsic phenomenon, however, this constancy is not entirely valid.

Since we are not familiar with phenomena of the ee domain—light possibly being the first—we are unaccustomed to think at that level. If we locate ourselves at the level of the electron, however, we would see light as something extremely small, so it is hardly surprising that, seen from the domain U, some of its characteristics are somewhat disconcerting.

The experiment of the pressure of light in an ampule almost evacuated to vacuum also indicates a loss of energy, and hence velocity, at constant mass.

We would also postulate the constancy of mass at the ee domain level since we understand, possibly intuitively, that mass cannot have planetary characteristics indefinitely when we go down to lower domains. As for mass/energy conversion, it is valid provided it is adequately interpreted. An example would be an atomic explosion: There is a loss of mass and a conversion into energy, as seen from U, but the disintegration of the elements (uranium, plutonium) does not mean annihilation of the mass. We should not forget that we have one mass in U, another in e, and another in ee. If the mass in U disappears, its elements continue in existence in e and ee; we have done away with the mass, and we call the product of this process *energy*, but this energy has a material substrate which was not destroyed.

If what we said about the physical basis of light—namely, a photon with quasitrons—is the case, then the velocity of the photon—*i.e.*, of light—is less than that of the quasitron. If this were not so, then the quasitron, which is circling about the photon, would be left behind and lost. If, on the other hand, the quasitron does not exist, this velocity greater than that of light would be that of the perturbation of the potential field produced by the photon and which is perpendicular to its path. This means that at the e domain level, there must be a velocity greater than that of light, both values being measured from the U domain.

In this regard, two things are evident: For material objects of the U domain, the speed of light is unattainable; and, as is said, in order to travel at that speed, an object would have to be converted into light. This is straightforward since, otherwise, how would objects from the U domain move in the potential fields of e? The only way would be by converting its constituents into e domain-type elements.

It is further said that a mass, if it were to achieve the velocity of light, would become infinite. In fact, this ought to be properly interpreted as the momentum being what becomes infinite, since what would have been added to the mass was velocity; hence the mass as such would not have changed. As for the infinity mentioned, it simply is not the case because the velocity of light, while high, is finite, and the product of finite values is also finite.

In mathematics, infinity is used for infinitely small and infinitely large cases, but in extrapolating this to physical reality, great care must be taken.

Now, if a mass were accelerated to the speed of light, its momentum value would be so large that its very existence would only be acceptable if the whole were to disintegrate into photons: If the mass of the photon is constant, then that of the source must also be constant; hence in the domains ee, e, and U, there would be no destruction of mass.

It could also be postulated that energy is movement and that if we look closely, going from domain to domain, the different aspects under which we detect energy—for example, in U—correspond to an aggregate of movements of mass in e, and so on.

In this way, we note a sequence of reasoning thus: If masses are constant and energy carriers; if this energy is movement, and movement, in turn, is the result of a force, then everything is reduced to forces; this is how many people would like to see matters.

It is logical to ask where the movement to be added comes from. Again, looking closely, we see that it is stored, as it were, in the domain immediately below that in question. Thus the origin of the primeval energy would be compatible with the initial "Big Bang" and its subsequent distribution and transformation in the universe U.

We realize, of course, that we cannot achieve total matter conversion into ee domain constituents: If we add an excess of energy to a substance it burns and we obtain fire, light, and finally ashes, mineral leftovers, or a new substance (for example, steels). Since something continues in existence, the conversion is incomplete and we are driven to state that in the ee domain mass continues as a constant value, even though it may have virutally no energy or momentum—that is, it has a velocity of almost zero.

Let us consider, as in Figure 25, that the photon is an entity, planetary in nature and detachable from the photon, electron, neutron, *et cetera*.

First, let us look at the background that allowed this conception of the matter, and the lack of which background evidently prevented such an advance being made in previous centuries.

Newtonian mechanics and Maxwell's electromagnetic field theory were the pillars of physics until the end of the nineteenth century. From then onward, however, there occurred a series of discoveries of phenomena quite inexplicable using those theories. Explanations put forward gave rise to a new set of hypotheses which crystallized over the years into what we know today as quantum mechanics. Max Planck was able to show clearly that classical dynamics, even in combination with rela-

70

tivity theory, is too narrow a framework to explain phenomena of the microcosmos, and he called for no attempts to seek more physical significance of the constant h—named after him—than that of the finite extension of the elemental region of physical space.

When Einstein compared the expression describing the entropy change, on a change of volume, of an electromagnetic radiation distributed according to Wien's law to the analogous expression for a system of particles, he concluded that a monochromatic radiation of frequency ν behaves as though it were composed of a finite number of localized and independent energy quanta with a magnitude given by the expression $E = \nu h$; these quanta were called *photons*. In other words, light of frequency ν is made up of photons with an energy of hν. Einstein then explained the photoelectric effect, in which part of the photon's energy is used to release an electron from its bond to a metal and a part is used in giving the electron kinetic energy. A result of this would be that the minimum kinetic energy is that which allows the release of an electron from the level of weakest bonding.

When Compton studied the collision of X or gamma rays with electrons, his explanation gave support to the corpuscular theory: he showed that on exposing a body to X rays, a secondary radiation, in addition to the dispersed rays, was observed. He went on to prove that the wavelength of the secondary radiation was independent of the target substance employed, but depended solely on the wavelength of the incident radiation and the corresponding angles of incidence and reflection. He also noted that the secondary radiation wavelength was always greater than or equal to that of the incident radiation.

In his explanation, Compton not only accepted that light of frequency ν is made up of photons with an energy E, but also that these particles have a moment P, where $P = h/\lambda$, and this moment was in the direction and sense of the ray of light.

According to the latest experimental evidence, we can calculate the mass of the photon using the expression

$$m_\gamma\ c^2 < 2 \times 10^{-21} \text{ MeV.}$$

In order to interpret the classical interference and diffraction experiments electromagnetic radiation had to be considered as waves, while the explanation of the photoelectric and Compton effect required the attributing of a particle nature to light. The answer to the question of whether light is a wave or a particle was only to come in the second half of the 1920s, mainly due to a statistical interpretation of quantum mechanics developed from Born's basic work, and the complementari-

71

ness principle, which associated each particle to a wave, announced by Bohr in 1927.

We now see that the matter may further be clarified by locating it in the domain in which it takes place. In order mainly to forget about the concept of a photon that sometimes behaves as a particle and sometimes as a wave, we understand the wave-particle duality, measured in U, and that this allows an interpretation, sometimes, as one or the other, but we have to accept that it would be complemented whether it continues in existence or not. If, in a phenomenon, we determine only a particle— that is, either the wave was not generated in the potential field through which it passed or remained undetected by our measuring instruments— and, inversely, if we detect, in domain U, a wave, what instrument do we have available to measure the mass of the photon?

It has been shown that photons have no mutual interaction, and they behave as indivisible particles.

Quantum mechanics was, of necessity, obliged to attempt an explanation of these phenomena and, since indeterminance is very large, to try to apply the deterministic criteria of classical mechanics. The following was done: A whole mechanics for the domain e was constructed, supposedly of an indeterminate nature. This was because it is difficult to determine the location in space of an electron at a given time, and this approach served to overcome many of the difficulties; as knowledge of quantum mechanics advances, however, there is no doubt that it becomes less and less indeterminate. This is the basis upon which the explanation of the nature of light rests—it exhibits complementary aspects of the same single reality, just as we said above. A further point is that ignorance about the domains is so great that it has been necessary to have recourse to taking variables, calling them hidden variables, until such time as their nature may be determined.

As for the difference between light and sound, we should note that sound is a simple mechanical perturbation in the U domain which certainly has an effect in the e domain level, but since it has something to do with elastic properties at that level, once the peturbation of the medium has passed by, everything returns to its original state.

We note too that sound does not propagate in space, as does light. In Figure 29 we see a diagram of a photon train with the revolving quasi-tron in motion (or the photon produced perturbation, if the quasitron does not exist), giving rise to the resultant transversal perturbation.

In Figure 30 we see another diagram of a beam of light formed by a photon train being reflected and refracted in a change of medium—in this case, that of an air/glass interface.

In Figure 31 we see a representation of the same situation as that shown in Figure 30, but in this case it has been taken that the quasitron does not exist.

72

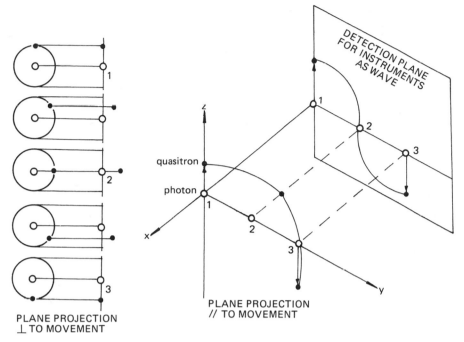

PLANE PROJECTION
⊥ TO MOVEMENT

quasitron

photon

PLANE PROJECTION
// TO MOVEMENT

DETECTION PLANE
FOR INSTRUMENTS
AS WAVE

Figure 28

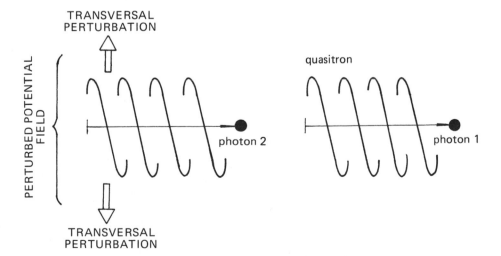

TRANSVERSAL
PERTURBATION

PERTURBED POTENTIAL
FIELD

photon 2

quasitron

photon 1

TRANSVERSAL
PERTURBATION

Figure 29

73

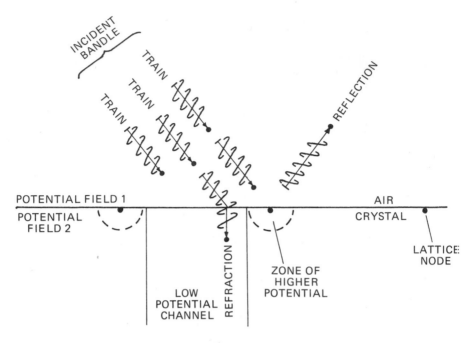

Figure 30

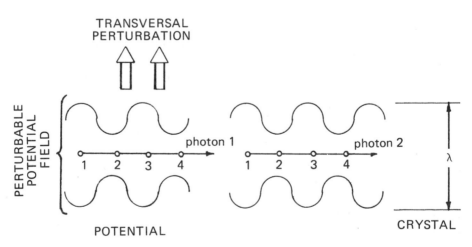

Figure 31

Chapter V

THE PROBLEM OF INFINITY; THE SHAPE OF THE UNIVERSE

The problem of infinities is a challenge to the intellect, to the imaginative capacity of man. It is, at best, very difficult to imagine how a positive and a negative infinity can be brought together, even using some trick or model to assist us. The fact is, nature has always shown itself to be simple enough; when it appears complicated, either we are on the wrong road, or nature is simply amusing itself by playing games. It is logical enough that there be a mathematics of large numbers, but one of infinites is difficult to conceive of.

Let us look at the problem: If we were, as Einstein imagined, to ride on a beam of light, sooner or later we should arrive at the limits of the universe U.

It has always been supposed that such a limit exists; hence in this essay, we have referred to what may lie beyond it as UU. This means that if U is finite, we have shifted whatever may be infinite into UU. This is the limit of our imagination at present: How many more than the four domains of today may we have to consider in the future as knowledge advances?

This again is difficult to imagine, and if we endeavor to think of something of infinite extent, it is simply impossible. On the face of it, it would appear that we consider the domain ee as being the limit, the bottom, the end of the road in that direction, because our imagination just does not conceive of anything smaller. We should not forget, however, that the dimensions of the photon belong to work done in this century. In previous centuries, before the microscope, it was only possible to imagine minuscule, dustlike particles and no more; even then, this was with the help of God.

Even so, we did say earlier that the idea of size only arises when matter is present—that is, dimensions, as such, depend on the presence of matter. In this sense, we can think of infinite space, since *infinite* means a very large measurement, and if we assume the absence of matter, there would be no dimensions, and that would be infinite. Maybe.

What, however, happens in the real world where we do have matter? We could say that the matter was put there, dimensions appeared, and that finiteness came about. This is fair enough for a region of the universe; what about other regions where there is more mass? This process has no end—the idea of infinity is beyond the limits of our minds; we could say, in fact, that it is bigger than our capacity to imagine.

In Figure 32 we see, in line with present knowledge, the domains represented as cubes. By means of our knowledge of the domain e, we imagine the other levels. At present, all the indications are that the ee domain is equal to those of e and U—*viz.*, planetary systems at the domain level following a certain body. Even so, we are unsure of anything regarding ee and UU.

We do, however, know that the e domain, as observed from U, takes on the aspect of solids, liquids, and gases, depending on the corresponding density or consistency. Would the ee domain be the same, seen from e? Would e seen from ee have the same aspect as does the universe for us, with its clusters, galaxies, *et cetera*? Maybe.

Now to the intriguing point: If we move down through domains, does everything end in ee? In another way, how does U look from UU? And what is UU anyway? We may try an answer based on our knowledge as described above—that from UU, the U domain would be like a solid, a liquid, or a gas, with life, living matter or mineral matter. Since we ourselves are here, we can say it contains life but at a very low level of organization. And how about the UU domain itself; what might it be, exactly? I have been bold enough to make a suggestion in Chapter VIII.

I can go no further than to imagine an infinite empty space, but one having some building blocks in it of the very smallest we can think of— for the moment, the masses of the ee domain. With these, by a succession of increasing steps performed upon a very special system of mechanics, other domains are constructed, on up to the largest.

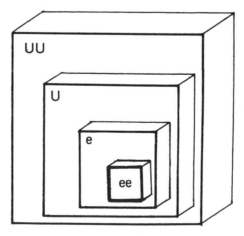

Figure 32

For such an exercise in imagination, I am constrained to forget about mass; and should it appear, then it too has to be in some way infinite for the whole to hang together. This, in turn, would only be possible if the basic unit of the mass—those building blocks—were to have a mass of zero.

Thus, if we were to dismiss the zero mass, but suppose that it can be very small, this would mean that space had never been infinite, but simply very big, in fact, in inverse proportion to just how small the smallest masses possible may be.

In conclusion, we have either a very large but finite universe with matter, or one that is infinite—if this means something—without matter. In this way, we arrive at the same point of departure by another route.

The only solution would be that nature is playing games with us, and everything is simpler than we think.

In fact, if we accept the criteria concerning living matter, stated in the corresponding chapter, we have to assume that the entities mentioned there are also in ignorance of the answer to this vast unknown. As far as our imagination is concerned, then, the result of their examination is beyond us, simply because it doesn't exist.

We can look back to some of the precedents on this matter and consider the ideas of the romantic universe makers. Giordano Bruno, above all, considered that the universe extended indefinitely in all directions, and that there existed stars and nebulae everywhere, just like those in the region of our Sun. It is said that if such were the case, then with the light coming from infinite luminous bodies, the sky would appear infinitely bright; and, most notable, that this infinite brightness would be, for the most part, produced by the more distant bodies. We think that if light expends energy, and its velocity is attenuated with time—though able to remain constant in local but large regions of U—and that light, in fact, does not illuminate as we saw, then maybe Giordano Bruno would have had a point. There is, of course, no account taken of light absorption by black holes.

In 1917, Einstein proposed a cylindrical universe in an effort to explain how extragalactic nebulae are able to maintain themselves in equilibrium and not fall inwards upon one another because of reciprocal attraction. This point may possibly be clarified somewhat in the light of the concepts put forward in Chapter II.

A further theory, proposed by Hubble, suggests a universe U of variable radius. There are other theoretical proposals, though to date none of them are definitive.

Chapter VI

BEAUTY AND THE ARTS

Works of art are probably the second great work of man and the longest lasting, if we place first those works that endow the human being with more mental or physical health. Among these last we can count religion, knowledge, or science in general.

Among works of art, I have a particular liking for music, and this I justify by understanding that the vibrations of music come into us, undoubtedly down to the e and ee domain levels. At those levels, there is presumably something that constitutes the basis of living matter, and this, in turn, "reflects" the vibrations back to the U domain, where they appear as—or give rise to—a sensation of pleasure. This pleasure or satisfaction must, at the e or ee levels, have an evident physical basis, since it is a physical phenomenon, which in those domains would certainly not sound like music; hence I understand it rather to be a response from those domains, and positive insofar as it would appear that we, in the U domain, become relaxed, quiet, and have other emotions, even tending towards those having something of a therapeutic nature.

We could reasonably ask what immortal work could science give to art today. Technology, as the practical application of science, brings us all manner of magnificent things, such as synthetic music, electronic instruments, *et cetera*. We also have very fast computers. Why not employ both of these elements to produce music? This brings us to the idea of an electronic composer—a computer programmed to create music.

Unfortunately, such machines would have no sentiments and would be unaware of the kind of music in favor at any particular time. We might try out what I call the idiot's method. Let's have a look at it.

First, we remember that at some point in times past, we were told that if all the letters of the alphabet were to be combined in all possible combinations in words and phrases, we would then obtain writings and books with all knowledge, both past and future. We also remember that the attempt is promptly abandoned upon realizing how many millions of years it would take to carry out such a project.

I don't think anyone is interested in this approach to obtaining knowledge over a series of generations, or, indeed, whether it would be convenient for the health of humanity, but it is possible someone may be. In the case of music, however, the situation is different. Music is healthy, and anything generated by the computer would be either good or readily recognized as not good and thrown out; we know that some results will be good.

First, let us make combinations of a set; we see that we can obtain certain results. If we do this with the numbers 1, 2, and 3, we get six results, namely, 123, 132, 213, 231, 312, and 321. Had we chosen to repeat the 2 with three elements we would have obtained 122, 212, and 221. Let us now do this with the letters C, O, S, and A, with no repetitions allowed. We get the following results:

1234	COSA	3124	scoa
1243	coas	3142	scao
1324	csoa	3214	soca
1342	csao	3241	soac
1423	CAOS	3412	SACO
1432	CASO	3421	saoc
2134	ocsa	4123	acos
2143	ocas	4132	acso
2314	osca	4213	aocs
2341	osac	4231	aosc
2413	oacs	4312	ASCO
2431	oasc	4321	asoc

(The combinations in capitals are the Spanish words for "thing," "chaos," "case," "sack," and "repugnance," respectively). We can see that in Spanish five combinations on twenty-four possible formulations are legitimate words—approximately 20 percent. This is the idiot's method, so called because, to obtain few results, a great deal of work must be done; today, computers can do a great deal of work. Let's develop this theme in connection with musical composition, assuming that the tremendous wait for works of high aesthetic value would be properly compensated by the resulting music. If this were, in fact, the case, we should consider those who did not try this method to be the idiots.

The point is that we want to get results, though not all possible results, since we don't know whether the universe would still be here by the time we finished going through all the combinations.

In the above Spanish language case, had we gone through only 50 percent of the possibilities, we would have had three meaningful results of the five possible.

If we try a similar exercise in English, we could select the letters S, E, A, and T.

1234	SEAT	3124	aset
1243	seta	3142	aste
1324	saet	3214	aest
1342	SATE	3241	aets

1423	stea	3412	atse
1432	stae	3421	ATES
2134	esat	4123	tsea
2143	esta	4132	tsae
2314	EAST	4213	tesa
2341	EATS	4231	TEAS
2413	etsa	4312	tase
2431	etas	4321	taes

It should be noted that in the program project to be followed, a sequence is supposed that may last so long that no computer would survive it, so two or more processors working in parallel and in series must be considered so as to duplicate sequences for greater safety and to ensure continuation of the processes should one of the machines cease to function for some reason. It should also be borne in mind that the project costs would be such as to allow only large companies to engage in the business, or possible governments with a view to some public benefit.

The basic approach would be to assign numbers or letters as codes to form musical scales; including tones and semitones, we would have fifty-six units. We would add fourteen more for seven tempos from semibreve down and the corresponding rests. In this example, we would have some seventy elements for combining.

Obviously, along with music of known types we would also get unknown types, among which would be some with no aesthetic value. There would be no golden rule for the resulting music, and we would expect everything from cacophony to harmony.

All we know is that what we are interested in will show up, and if we think that at some moment we will hear the chords of Beethoven's Fifth Symphony, then the effort I propose would be worthwhile, because we don't know if there will be another Beethoven.

Thus, if t_1 is the time for a processor to synthesize, and t_2 is that for an instrument to play each partitura, then we could conceive of a music synthesizer as shown in Figure 33.

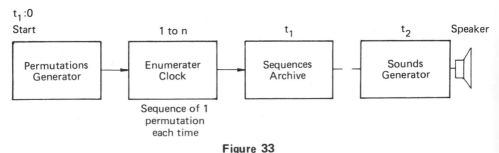

Figure 33

80

The choice of which music is good, of course is for the human beings who would have to stand permanent guard, in shifts, next to the apparatus.

In order to follow a logical sequence (with an alogorithm) the following tree could be constructed, where

n = sound elements for repetition
m = number of repetitions (length of the partitura).
Example: n = 70; m = 100, and so $n^m = 70^{100}$

A screening unit could be incorporated to reject obviously unacceptable combinations, such as a series of equal terms, disharmonies, chords not forming exact tempos, *et cetera*. If we assume a 50 percent rejection in this way, we would have

$t_2 = 70^{50} = 1.8 \times 10^{92}$ seconds $= 10^{83}$ years (at 50 percent: 10^{42} years).

Even supposing we were to achieve a one thousand-fold speed increase in this process, the 10^{83} figure is reduced only to 10^{80}.

In spite of these difficulties, the criteria for going ahead have already been mentioned. Some may even wish to go ahead because of the difficulties stated, considering it inconvenient to do without them only to face a rapid exhaustion of the aesthetic possibilities of music.

I have set forth below an example of a permutation, or combination, generator.

Data: Series
    1) Codify in binary from 0 to 70 (sign allocations are for the case of results print-out and series input)
    2) Make tree of 100 columns of 70 elements each (70^{100})

Logic:

A 1) Store series 1 to 70 in memory 1. Boxes 1 to 70
A 2) Store series 1 to 70 in memory 2. Boxes 1 to 140
— —

A100) Store series 1 to 70 in memory 100. Box 1 to (70×2^{99})
      Take and form the results of:

B 1) Memory 1. Box 1 transfer to memory 2, Box 1 ... memory 70, box 1.
B 2) Memory 1. Box 1 transfer to memory 2, box 2 ... memory 70, box 1.

B 3) Memory 1. Box 1 transfer to memory 2, box 2 ... memory 70, box 70, etc.

n = Column No.   2   4   8   16
                  $a2^0$ $a2^1$ $a2^2$ $a2^3$

a = 70 (No. of boxes)

$70.2^{(n-1)}$ Locations   Locations
                         1 to 70   1 to 140

We have already observed that the human imaginative capacity must have a limit, hence a method such as this may well serve to broaden our intellect, something possible today with the aid of the computer.

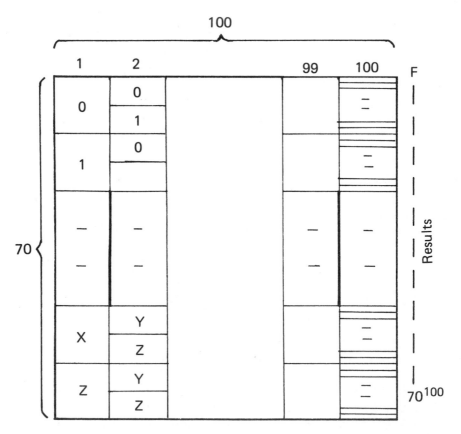

Figure 34

Memory 1	Memory 2	Memory 3	Results
1	1	1	1 1 1
		2	1 1 2
	2	1	1 2 1
		2	1 2 2
2	1	1	2 1 1
		2	2 1 2
	2	1	2 2 1
		2	2 2 2

Figure 35

We can suggest a scientific method consisting of a sampling of all the possibilities for a phenomenon to occur, determine its value, make transformations, *et cetera*, such that the computer, by reviewing all its files, could guide us toward a solution for particular problems. This would appear to be how the human mind works at the unconscious level, giving rise to solutions appearing, occasionally, only after a distinct delay. We are fortunately unaware of the processing described, otherwise we may have some difficulty in using our brains on the conscious level for anything else.

Example: $2^3 = 8$

Using the following table code and combining these 44 elements with repetition taken 25 at a time gives 14×10^{11} number of combinations.

DOW1	DOB1	DOC1	X	Y
REW1	REB1	REC1		
MIW1	MIB1	MIC1		
FAW1	FAB1	FAC1		

83

			X	Y
SOW1	SOB1	SOC1		
RAW1	RAB1	RAC1		
SIW1	SIB1	SIC1		
DOW2	DOB2	DOC2		
REW2	REB2	REC2		
MIW2	MIB2	MIC2		
FAW2	FAB2	FAC2		
SOW2	SOB2	SOC2		
RAW2	RAB2	RAC2		
SIW2	SIB2	SIC2		

With a fortran program it is possible to make trials with different inputs aiming to a fast finding of the 10 first measures of the 5th Symphony of Beethoven:

```
68136   YSOC1SOC1SOC1XMIW1YFAC1 FAC1FAC1REW1REW1YSOC1SOC1SOC1
        XRAC1XRAC1XRAC1YXMIC2XM IC2XMIC2DOB2YYDOW2

68137   YSOC1SOC1SOC1XMIW1YFAC1 FAC1FAC1REW1REW1YSOC1SOC1SOC1
        XRAC1XRAC1XRAC1YXMIC2XM IC2XMIC2DOW2DOB2YY

        BEETHOVEN WAS FOUND !!
```

In this case it was found in the 68137 calculation after about 1 minute with a Vax System.

Chapter VII

LIVING MATTER

INTRODUCTION

In the area of frontiers in the biological sciences, we shall first summarize some of the tendencies before trying to present some ideas in connection with the quantification of those sciences. First, we ought to forget the concept held by some biologists in the sense that the era of biology is about to open, and that it—biology—will surpass physics. We also have to remember that we are not really in a race and that the advances in biology are quite apparent. Even so, the moment we speak of the real world, advances in biology depend on physics; the opposite is, while not impossible, at very best a rare and isolated occurrence. Even then, the matter is not one of relative inferiority; there is simply no sense in this approach.

In the light of the scientific advances of the last hundred years, mainly in the area of atomic physics, we can reasonably claim to be unburdened by traditions, laws, ideologies, religions, *et cetera*, which, in the past, endeavored to impose a particular rigidity of thinking patterns such as to render acceptable their corresponding descriptions of natural phenomena. The situation is such that, in research, the important thing to remember is to be prepared for the unexpected.

Owing to the inappropriate use made of scientific advances by those in power—governments—it has become necessary to strive for a moral system in keeping with the scientific era. Unfortunately, the powers capable of producing such an updating process have themselves remained far behind in relation to new discoveries. The scientist's contribution is but a reflection of intellectual and objective integrity, tolerance, acknowledgment of error, and suchlike. The support that physics can now offer to explain the phenomena of living matter will necessarily include a consideration of time, a parameter to date given little significance in biology. We saw earlier that on shifting from domain U to domain e, a time-scale change must be made. Now, looking at something which, observed from U, appears as life, the above-mentioned scale change must not be forgotten.

The moment we speak of specific energy states encountered, at any level, in the form of masses in movement, we must conclude that in systems of any domain complex and dynamic in nature, time will play an important role—and this *time* will not be *absolute time*.

Curious though it may seem, we should note that the most spectacular advances in the past 150 years have been in the areas of astrophysics, fundamental particles, and biology.

The idea of a self-regulated and regulating universe became popular, but a proper understanding of how such a mechanism might work has not yet been achieved. The hope behind some sort of unification of physics and biology is that gaining an understanding of how this might come about may give man a way of understanding himself. This is to say, he may be able to find a physical basis for living matter, thus to be able to explain life-associated phenomena that, up to now, appear to be divorced from physics altogether. There are those—psychologists and suchlike, for instance—who would keep this field apart, not allowing a global view of phenomena.

This, of course, brings us to the difficult question of what life and consciousness are anyway. In how far can the different parts of a living being be replaced and still allow for speaking of the same individual? The limit seems to be the brain; but then, we ask, what is the brain? There is, in fact, no specific group of master cells that guide and direct phychic activity: The brain is a group activity, where the overall whole of chemical reactions, with their bifurcations and complexity, probably constitutes the individual.

Within living systems or living matter there are extraordinarily strong laws, which, although emanating from the microcosmos, are strong even in comparison with cosmic forces. Here again we must not confuse cause and effect: It is not that the forces associated with these laws arise because of the particular system, but that the system exists because it is there to make manifest the underlying forces; otherwise it would not exist.

In all of biology there are aberrations and behavior that would appear to be contrary to its best interests. This, at first glance, appears as something negative, but it allows living matter to look for evolution routes in line with the tendency of Innovation—Deviation—Tendency, and New Form. The case of cancer is a paradoxical one, since it cannot be claimed that a cancer cell has learned somehow to reproduce itself without dying, or to have found a road to immortality, or that it is immortal; in the end it kills its host and finally dies itself.

Advances in biology went along the cell, microbe, molecular, and ionic routes, and today it is on the threshold of particles at the atomic level, just as happened in the case of physics.

Returning now to the subject of the base or seat of phychic activity, the brain, we must make a distinction between the paleocortex—or primitive part of the brain where instincts, endocrine control, and everything that contributes to immediate survival are located—and the neocortex, made up of a series of layers of neurons that have accumulated over the last two million years and that has a capacity for perspective, symbolic thought, and self-expression. There is an equilibrium between these parts of the brain which forms a complex unit with many thousands of feedback, regulatory, and other systems.

Moving on to individual persons, we can say that their reactions are made with, and in the light of, the baggage they carry of all previous experience and learning. This is made up of all they have stored in their brains—all their opinions, prejudices, and everything they have seen as good, evil, attractive, and ugly. In the nervous system there are some coded routes that conflict, in the expression of action, with other routes that are coded otherwise. The only way these phenomena take on conscious form is in expression in language.

Biologists believe that the greatest opening for the progress of their science in the coming fifty years will be a knowledge of the information-handling mechanism of the cell. This is something fundamental, and it goes beyond the problems of health and disease. In the living cell—whether bacterial, vegetable, animal, or human in origin—in that part of it where energy change takes place, there can be seen an energy-transformation entity capable of transforming one kind of energy into another, if metabolizing food, producing waste for subsequent elimination, and so on. This whole affair is contained within dimensional limits that are, for us, difficult even to imagine, and even more so if we think of the changes of scale necessary to arrive at a complex comprehension of such phenomena by means of a comparison.

In the cells themselves, there are two aspects: energy interchange and information handling. This latter, for example, allows the cell, in cases of allergy and immunology, to retrieve information from the memory and shift it to its immunological consciousness. In the final analysis, the cell's capacity to recognize a chemical substance as friend or foe—known as the psychological or physiological memory—is a consequence of its information-handling aptitude.

As for energy transformation, the adenosine triphosphate (ATP) energy-generating system is more readily visible for us because it is several thousand times larger than the information-handling system inside the cell.

If we take the relative sizes of a processor and its battery as an example, we see that the battery (the energy-generating system) is several tens of thousands of times larger than the processor. Since the cell's capacity and efficiency are far superior to any processor made by man, what order of dimension could the cellular processor have? Whatever those dimensions may be, the unit is, without a doubt, minuscule.

For the progress of biology, the comprehension and control of the information-handling system is the important thing; the energy aspects slip quietly into the background.

A BASIC POSTULATE

In previous Chapters, we became accustomed to surprises on looking into events in the domains below our own, so we now can reasonably postulate the existence, at the e and/or ee level in living matter, of entities or beings with a certain independence and reason now definable in terms of its properties.

This brings us to a consideration of matter as mineral, organic, or living organism.

In mineral matter, there would be no beings such as those mentioned above. Organic matter would form a basis for the existence of entities, but their existence is not presupposed. In living matter they would be present, and in a living organism, the living matter becomes autonomous and can reproduce.

There is an (unjustified) temptation to see such entities somehow as minipeople.

Examples of the above categories would be sodium chloride, oil, fresh meat, and plants or animals.

So it is that we come to the concept of the equality of worlds—in other words, the assumption of an equality of phenomena in all the domains. We would consider it a fruitful approach in the physical sciences, and the indications of experimental evidence are that it is a virtual certainty. The other point, of course, is that we have nothing better to propose. In this vein, then, it would be worthwhile trying to see what results it gives in regard to living matter and life overall.

If we ask ourselves who put the beings there, we can expect an answer when we find out who put matter in space and where it came from in the first place. This being the case, we could try the thought-experiment approach and ask what we would see if we were to gradually reduce our size such to be able to stand in the vicinity of an atom in organic matter and have the same size in relation to an electron as we normally do in relation to the earth. We would doubtless see a universe of great complexity, symmetry, and so on, because organic molecules are very large and complex.

If we now return and make the same trip in simple living matter—for example, to a molecule of a liver cell—we would see the same domain (e) and complexity as before, but with certain beings living on the surface of those small planets and forming part of a mechanism, visible by telescope from one of them, of atom interchange in other molecules passing by. Such beings would form a part of the mechanism receiving information about what, when, and how to proceed for the reception of atoms arriving in their region. This would signify that metabolic activity with its chemical messengers could be explained.

The beings themselves—their lifetime and ongoing existence—would, of course, be subject to the circumstances of the U domain being of which they form a part, and they are aware of it. This is the why of their struggle and perseverence in their efforts to comply with their commitments. To imagine the existence of the beings is not difficult for anyone who has observed real living things visible only through an electron microscope, and if these are made of tissue—also living—and the tissue, in turn, of molecules, then straight off we would consider that in the molecules of that living tissue there is something, and it is up to us to find out the properties of that something—namely, the beings spoken of above. It has been said that living matter is matter in its most complex form. If such be the case, then we begin to wonder whether these beings were what brought about this organization, or was it formed at random or by chance in its molecular complexity, and the beings "moved in" *a posteriori*. This becomes a chicken-egg question. It is apparent that some molecules in some circumstances may give rise to organic matter, such as in the case of the primeval organic soup. This matter, however, is organic only—not living. One conclusion would be that on a complex organic substrata, in certain circumstances there may "appear" living matter, explained by such beings as those mentioned above.

We ourselves, after all—if our galaxies were part of something considered living matter, and ourselves a part of a system that survived by receiving energy—would carry out our tasks to maintain the status quo; anything else would mean suicide.

Let us now move to a living being and down to domain e or ee, where we would expect to find our "beings." This time we choose the nervous system in the brain and we endow the "beings" with the power of speech, thus to hear their story. It might go something like this:

"As far as we are aware, we live in this universe on these planets circling the sun you can see over there. We depend on the energy arriving in our system, and our technology has been such as to allow us to construct this magnificent civilization, although there are still a great many things we don't know. For instance, we don't know what may be way out there in greater universes, or, indeed, what may be way down there in lesser ones. What we do know is that if we observe the organizational laws of our own universe, we can obtain what we need to survive. We know we will die sooner or later, but our instincts tell us that there is a higher power which makes us somehow immortal, in some way, as though the greater universe were reproduced in greater beings; hence we believe in immortality. We also know of and are in contact with similar civilizations which ask us for information, which we supply. We too request information, which we process, and we send it to where we are asked to do so. We believe we are fulfilling an organizational function, and we know that other regions of our universe carry out other functions—the provision of energy for instance. We also know that a part of our universe specializes in obtaining information from beyond its limits, and this we receive in the form of vibrations, which are of assistance in our overall struggle for survival. We also make up part of a number of planetary systems in our local region, and, having solved the problems of energy supply and thus enjoy a material surplus, we have time to discuss among ourselves those problems that are the basis of knowledge. So it is that we exchange information as to who we are, what we would look like from a higher-level universe, and suchlike.

Well, all that would remain to tell the beings—since we are in their higher-level universe U and that we observe the e or ee domains—is that we conceive of them as something akin to a soul, and the information they exchange between themselves we call *thoughts*. We could go on to say that their fears of perishing for lack of energy supply would correspond to the death of a being which at the higher level of its own universe is reproducible; they need not be so apprehensive of perishing insofar as there is method of propagation that is infinite in time; the vibrations they receive originate in universes at their own level highly specialized in analyzing similar vibrations originating in higher-level universes, *et cetera*.

I am not going to try and define either *organic* or *living* material. I do wish to make mention of there being many forms of life, from the very elementary up to the very complex. Again, it has been said that there could be life with entirely different characteristics—for instance, electric life, the medium of which would be conductors of electricity, the world that would be known to the inhabitants would be that of metals, electrolytes, *et cetera*—and one would mention all manner of similar suppositions. Since, however, we want to solve our own enigma, we can most fruitfully concentrate on what is similar to us. I do not wish to spend time either on the organizational aspects, which are long and well treated elsewhere.

I would like to note the approximation that I discern for justifying the biological progress of the human being. This is an area in which one must be on the lookout so as not to confuse cause and effect.

It is, for example, clear that sex has played a major role both in evolution and psychology—far beyond its function as an instrument of reproduction. We are faced with the question of whether the difference between man and the animals is a product of their evolution, or the evolution is a product of the sexual phenomenon. I believe that man,

90

the more he evolves, seeks more pleasure, both in quantity and quality, and among these pleasures must be counted those of the intellectual variety. It might be mentioned in passing that the year-round sexual activity of man is a consequence of his having evolved more, rather than less.

There are many other matters that can profitably be looked at in the same fashion—the confusion of cause and effect is common in the on-going search for explanations. The effect must be clearly identified, or we must follow through on a series of variations until we confront contradictions such that we are obliged to adopt the contrary as true.

We have seen that by postulating living beings at the submicro level in organic material, whose purpose would apparently be to maintain life for successive generations in that medium, we can arrive at a definition of living matter. Let us suppose that organic matter forms adequate substrata for the appearance of the beings and we adopt an intermediate position regarding the choice of the cause of organic matter becoming living matter—which is reasonable, because organic matter provides adequate conditions for their proliferation.

If this were the case, and we grant such beings some capacity of collaboration in the organization of their worlds, then they must also be given some degree of intelligence. Going further, this intelligence, seen from domain U, appears as what we call the animate reactions of living beings, among which, in the first instance, would be those of vegetable life. These would be those reactions that have become more or less automatic because of their continued repetition throughout the time of the evolution of living beings. It would be easier to spot such reactions in plants than in animals, since plants have suffered relatively less evolution. They must also be apparent, of course, in the nonautomatic higher beings—nonautomatic in the sense that some reasoning is required in order to act in an appropriate manner. This power of reason we should call *intelligence*. Since we have reactions at both higher and plant levels, and intelligence at the higher level is manifestly present, then should we not also have a—very limited—intelligence at the level of plants? I think so, and this broaches the second aspect to be borne in mind if we wish to interpret biological phenomena quanta-tively (in the sense of quantum physics, or the study of the matter on the basis of a minimal discreteness), which is that of assuming the conceptual unity of living matter, adequately interpreting the different facets thereof in the different beings. If human beings have specialized tissue where intelligent reasoning activity is mainly localized, should not plants also have something similar? Surely, but then where would we find the brain of a plant? We would never expect to find anything large and tangible; rather, we would suppose that their tiny rational activity would be located in simple tissues or cells very small in size. The activity itself has been detected, mainly because virtually the only way open to plants to manifest such is by visible physical changes—chemical reactions, growth, and suchlike. There are, for instance, reports of greater growth in the presence of musical sounds, chemical markers at ground level and vital for the plant's survival, and changes in the coloration of flowers to facilitate polination, among others. All this points to there being a tiny brain in there, if only it could be found—a job for botanists, or physico-botanists of the future.

In the same vein, we would be constrained to consider that animals have greater reasoning power than plants, but less than man. If then we accept what happens in the living matter of man at the level of domain ee and that such is also the case for animals—but not with submicro beings at a lower level of evolution—then we would be left with man as constituting their highest achievement. If we were to take man's soul as a manifestation of life at the ee level, then animals should also have something of a soul. What this may be like, of course, we don't know, though animal lovers, at least, are convinced that it is there, and anyone watching a dog whine, or seeing other emotional reactions in higher animals, tends to become convinced too. If the ruling norm for man is supposed to be "love others as one loves oneself," then we are, as it were, from the same manufacturer, and we should accord to animals consideration, respect, care, *et cetera*—not to the same degree as they do to us, which would be a straight exchange, but in acknowledgment of their constituting beings fundamentally of the same nature as ourselves. If it were the case that the lesser evolution of animals relative to humans were due not to lesser evolution at the ee level—but rather it's the same at the level and was just not brought to the same plane of development at the U domain level—then we would have to conclude that at the ee level, with their beings, the animals are the same as ourselves.

I do not want to finalize my comments on this matter without at least mentioning some of the most prominent figures in this century whose pioneering imaginative concepts of the microcosmos, later confirmed by experiment, have given rise to the greatest advances in science and allowed us today to enter into those worlds.

Among those men of science I recollect Niels Bohr, de Broglie, A.H. Compton, Paul Dirac, Enrico Fermi, Werner Hersenberg, Ishiwara, Hendrik Lorentz, Max Planck, Ernest Rutherford, Erwin Schrödinger, and Sommerfeld, to name but a few.

With regard to other areas relating to psychology, we might mention:

(a) in connection with brain waves, telepathy, which is disconcerting insofar as it is undetectable by any instrument other than one which is the same as its transmitter—namely, a human brain;

(b) Transmigration of souls (This is a self-evident phenomenon, which brings to mind a belief from ancient Greece: It was thought that by consuming greens grown in the holy field, one would receive the soul of the dead—a little hairy perhaps, but it is not the only mention of a supposed physical passage between ee domains. An interesting question would be that of the feasibility of a successful transplant of a single ee level being—or *monad*, as Leibnitz said.);

(c) in regard to psychokinetic phenomena, the movement of objects by mental power (We could posit the possibility of waves from ee level in the mover acting upon the e domain in the object moved. The matter of levitation raises a problem, as it were, for those who have seen it, insofar as it concerns heavy bodies being kept up in the air.);

(d) ghosts (Could this phenomenon be explained by the movement somehow of living ee domain beings from that domain of the matter involved in the events?);

(e) spirits (This phenomenon looks rather like disembodied souls, and it may readily be thought of as having no physical basis. If, however, we think in terms of domains, we might consider some e or ee level material or a contribution thereof acting as a base for the beings able to survive and manifest themselves in the U domain, which is where we ourselves are sensitive.).

As for memory, having already noted the similarity between phenomena in the different domains and the matter of transferring our ignorance of one—in this case, U—to another—say, ee—we might suggest that it is a filing system on those lower levels, something like the history of the beings recorded in their domains, in line with the information available to them from higher domains.

We know that inside the nervous system there are intercellular connections which function by means of substances known as neurotransmitters. We could thus consider that aging and the concomitant loss of memory was associated with the exhausting of some neurotransmitters.

We should mention, of course, that at the e level, and seen from U, we call these things *chemical substances*. In the light of the criteria given, the reader may like to look into the options for a physical interpretation of the phenomenon of dreaming.

Chapter VIII

LIFE IN THE DOMAINS;
THEOLOGY

With the introduction of the beings in Chapter VII, there is a temptation to consider that they and the soul are one and the same thing. This is to say that, from the U domain, the effect of their activities is that which would normally be associated with the soul of a person.

This brings us to the question of God and in the scientific world there is a widespread belief that if God didn't exist, it would be necessary to invent him. The reason for this is the known human requirement to explain the inexplicable.

In the last couple of centuries it was commonly held that with the surprising advances in science, it was no longer necessary to believe in God and that there were other rational explanations for everything; strong atheistic trends appeared. All this notwithstanding, it could be seen that with rapid and continually surpassed scientific progress, there was still no knowledge of nature sufficiently precise to allow a shift from the concept of the divine to some definitive, rational one.

At present, as the frontiers of knowledge are pushed farther and wider, there is still no inkling of a satisfactory explanation of facts and events; there is a return to the idea—not of the existence of God—but to the necessity that such a belief will continue.

Let us consider why this is so. Formerly, primitive man needed belief to explain the mysteries of his existence, birth, purpose, and death, among other things. Civilized man, with his culture, also needs belief to justify the incomprehensible, although it may be on a higher level than before. The intellectual who sees, as we have seen in some measure here, the constant widening of horizons without there being a "final" explanation of anything also needs belief, although he doesn't justify it and rather hopes that the need, at worst, is a temporary one.

What then will be the situation when, having finally disproved the existence of God in the domain UU, we find there is a UUU and an eee?

If things turn out so, then God is likely to be around for a while yet; if not, then praise God, we will have the answer.

Considering the time scale in this matter, let us look at when we may have a thoroughgoing knowledge of UU- and ee-based phenomena—in one or two millenia, perhaps. Anyhow, it will be a long time. What should we suggest for the intervening years? Can we be spokesmen for atheism when we are aware that there is no answer to the God-and-life questions and there is not likely to be for whole eons? This would but deprive whole generations of the race of consolation, a sense to their lives, and the belief that incomprehensible nature events are somehow justified.

This is obviously no road to take, and it is for this reason that I believe the scientific community should take advantage of every opportunity to denounce atheism as a crime against humanity.

This is not to recommend the other extreme; excess in religiosity is also pernicious, as also are those who fail to fulfill their own functions, such as the religious who dabble in politics and politicians who involve themselves in matters sacred.

Throughout history, God has been variously located at or in the Sun, personified in Greek and Roman deities, or identified as the One (God), *et cetera*. Given present knowledge, where should we locate him with a view to identifying Him by the effect He may have?

I think he belongs in the UU domain; moreover, I would think the UU domain *is* God, insofar as if we apply the comparison of ourselves through the results of our activities, being a sort of God for the beings down there in domains e and/or ee, then the body, mass, matter, or being that is UU could, because of its effects, power, virtues, wisdom, and as the virtually limitless source and provider of our needs (energy) be a God for us.

It also seems logical that if the soul is to be located in e or ee, the place for the Almighty ought to be the macrocosmos.

Having read justifications for the existence of God by little more than plays on words by men of the stature of Plato, Descartes, Newton, and a whole string of others, there is a certain satisfaction in these efforts to justify God's existence from something of a more realistic approach, even though it is entirely without supporting evidence.

Nonetheless, it is more in keeping with the scientific spirit to put off a solution instead of to promptly disqualify a realistic demonstration using a logic based on ideal concepts.

Chapter IX

A CONCEPTUAL SYSTEM

Let us postulate the similarity of phenomena in all the domains, with the possible observation that things become a little more complex from the ee domain through to the UU domain.

We would further assume that there is some form of life in living matter down at the ee level, and hence, until such time as we know the increase in complexity is reasonable, we would say that this form of life has the same characteristics as does life in U.

Concerning the overall organization of the universe, considering all the domains, we can gather that the conservation of momentum principle is valid, as far as we know, in local regions of the universe, and also that this principle is transferred between domains.

What then is going on in regard to the thermal degradation of the universe? It has to do with the energy conservation principle; we could say that, with the passage of time, energy has been flowing from ee to U. Thus it may be supposed that there was a primeval energy that started everything and originally came from the UU domain.

Another question is that the radioactive material. We may suggest that in pre-Big Bang "time" these substances were formed either by addition of force from the "outside"—as we mentioned previously—or in the agglomeration of mass itself into a compact unit which subsequently exploded, generating the universe we observe at a certain stage of this Big Bang, as though in slow motion; radioactive substances may be left-overs of that primeval material that continues to readjust itself at the e domain level and perhaps also at the ee level. If such were so, then we may think that with evolution of the universe, we may sooner or later get to some limit of expansion, with a shift from the transfer of momentum to the domains back to a new concentration into a monolithic unit of matter giving rise to a repetition of the whole affair. It would appear that nature abhors repose. In this case, the explanation of such a reversal would be readily accessible using the concept, set down in Chapter II, about the nature of gravity whereby the action is located in the moving mass and not in that which acts as the central mass.

If there were a linear transfer of energy from UU to ee, we would have to consider the possibility of energy annihilation. Energy annihilation goes against the grain; we have to consider energy transformation but not energy destruction. We take the apparent disappearance of energy—such as wear and friction losses—to be deceptive and that such losses, observed from the next domain down, are no longer losses. If we look at friction in the U domain it appears as an energy loss—at best a transformation which over all U would finally lead to the energy death of the universe. At the e level, however, we obtain momentum from the shifting of rotating masses and movements in empty space to positions of greater potential energy with respect to the nucleus which acts as the central mass. Where then is our loss? If we tot up everything from the ee level, we may begin to ask whether the so-called energy or heat death exists at all and whether we are, from the energy standpoint, in a pulsating multi-universe. At the present time, this is something complex and rather more for future workers than those of today. We can, however, say that a uniform distribution of masses in space would make U, observed from the UU level, appear as crystals in e look to us from the U domain.

In this system of ideas based mainly on the concepts that have emerged in the last hundred years, it has been assumed that since we have a macro- and microcosmos, it must be possible to pass from one to another if we use adequate transformations for domain passage. This is an intellectual exercise and I am not aware of it having been done before in this way but I do feel it is an approach that may be fruitful if worked upon with some precision.

It may be that, to date, the only domain passage constant for a shift from U to e is that of Planck. There must be others, and I think it is most important to arrive at domain passage formulae for the basic dimensions like distance, inert and weighting mass, forces, potential and inertial fields, times, equivalent types of geometries, for the demonstration of the validity of classical and relativistic mechanics in the different domains, et cetera. Undoubtedly, only the progress of knowledge will allow for significant experimental progress. Even so, the way must be prepared so that this preparation can suggest and give rise to the particular experiments to carry out so as finally to permit us to complete, with concrete data, the overall conceptual edifice.

The indetermination and randomness of quantum mechanics will thus begin to give way to the determinism and causality that we seek, although the process will be a long one. Thus we can arrive at an expectation of the unification of science, starting, in physics, with the unification of the fields, something that, as we saw, appears to be feasible. Moreover, this unification is in line with the way nature works and is not an artificial human aspiration.

Having gone through a rapid summary analysis of how the universe, as a whole, works and the functional relationships apparent between the macro- and microcosmos, on arriving at the present frontiers of physics, we can look at the prospects of physics in the coming centuries and the aspects of renewed physics.

97

First of all, we can expect that in the direction of the e and ee domains there will be an increasing determinism and that such progress can and must arise out of theoretical schemes and new experiments parallel with such schemes, facilitating their refinement. The other way, of course, is that of virtually accidental and/or isolated or particularly ingenious experiments by some workers which may help us along in the same general direction.

In the U and UU domains it is apparent that our only option is astronomy and space exploration to add to our knowledge, as did the locating of an infrared telescope in orbit which gave results of a kind previously unexpected. If infrared radiation is from the e or ee domain, then in this case we see a triple collaboration of "effort" between physics, astronomy, and technology.

We can, today, reasonably suppose that the advances in the e and ee domains will be of assistance in the less accessible and far-flung regions of U and even for a glimpse into UU.

In matters biological, everything depends on the instruments that physicists come up with to investigate the e and ee domains; such instruments will facilitate the clarification of the enigmas of life at the submolecular level. Even so, in view of genetic engineering and other similar advances, it can be expected that information about the e and ee domain levels in living matter will become available by indirect routes.

In regard to the suggestions made here, we must check to see whether they fulfill the requirements of acceptability as scientific guidelines, which would be (1) that they assist in the interpretation of the unknown in terms of the known, relating independent facts in a logical and readily accessible mental scheme; and (2) that they suggest new relationships to us, facilitating a unifying framework between the old and the new facts and widening horizons.

There ought also to be some things suggestive of how to observe new phenomena and to solve practical problems in a context of expanding ideas and setting limits to or replacing old ones when necessary.

Einstein stated that the only justification for our concepts is that they serve to represent all the complexity of our experience. He also said that there is no logical way to discover elemental laws there is only the route of intuition assisted by the feeling that, beyond appearances, there exists a certain order.

Chapter X

THE PHYSICAL AND PHILOSOPHICAL ANTECEDENTS

PHILOSOPHICAL ANTECEDENTS

We can now turn and see how the postulations made here, with the benefit of the advances of science, have in fact been implicit in the thinking of many physicists and philosophers over the centuries.

We can start with the philosophers. We know that many different schools have flourished through history, such as rationalism, vitalism, nihilism, and suchlike. We can look first at Aristotle.

According to Aristotle, Melisus stated that if something had existence, then it also had to be eternal, since nothing could come from nothing. This was so because, whether all things or only some things had come into being, it was clear that in both cases they are eternal, otherwise things having existence would have come from nothing.

He also says that a void is inconceivable insofar as a void is nothing, and that which is nothing cannot exist nor move nor displace itself to any point but that it be full. Thus if the void were to exist, it could withdraw into the void; hence, since it doesn't exist, it obviously has no place to go.

The philosopher Heracleitus makes mention of the flux of tangible things, the continual change, and that without ongoing change or alteration, there would, perhaps, be no such thing as experience. Any paralysis would be inconceivable because everything in existence is in transit in space and time, which would mean that all things never cease or finish their task. No one travels on or crosses the same river twice because the waters are always other than on previous occasions.

If the world is in an eternal state of flux, there is no need to look for anything permanent. Man is the same as nature, being something that is constantly integrating and disintegrating.

Wisdom is a single whole, with its purpose being to know the reason why all things are guided by all things.

How is it possible that the soul, having "lodged" with the Divinity and lived the most noble existence, can descend and enter a body?

According to Aristotle, everything living has some kind of soul. The vegetable soul, possessed by all plants and animals, only involves itself with growth, while the animal soul is the cause of the movements of the body.

99

Let us look now to Socrates and Plato: All souls are immortal since everything that moves of itself is immortal. That which moves another or is moved by another, on cessation of the movement suffers a cessation of life. Thus only that which moves of itself, provided it does not abandon itself, never ceases to move, and also is for all things that move of themselves the source and principle of movement. The principle, however, is unbegotten because it is necessary that everything be born of the principle, and this is born of nothing at all. Furthermore, since it is unbegotten, it is necessary too that it should be imperishable inasmuch as if the principle were to perish, it cannot be born again of anything at all, nor could anything be born of it because it is necessary that all shall be born of the principle. Thus it is that the principle of movement is what moves of itself. The death or birth of this principle is impossible, otherwise the whole universe and the process of generation would fall apart and come to a halt and would never again have a source from which to arise, with movement, into existence. Having shown that that which moves or itself is immortal, there would be no shame in stating that precisely that is the essence and the notion of the soul. This is so since all bodies receiving movement from outside of themselves are inanimate and all which receives its movement from inside, from itself, is animate, as though this of itself encapsulated the nature of the soul.

If this is the case—viz., that all that moves of itself is soul—then, of necessity, the soul would have to be something unbegotten and immortal.

Continuing with the call for assistance to ancient and modern thinkers to support our proposals, we are acknowledging them, too, with postulates that corroborate their lines of thought.

Moving on somewhat, we come to Descartes, who said, in principle, that all men were equally endowed for arriving at the truth since all have the faculty of reason, which allows of distinguishing between the true and the false. Human fallibility, however, is always a threat to men's conclusions; hence, a method is required.

Descartes's fundamental thesis is based on the principle *cogito ergo sum*—I think, hence I am—and his proofs for the existence of God are based on the impossibility of thinking of the imperfect, without that this should presuppose the perfection of the Divinity.

As for knowledge, he says that there are but two points to consider: ourselves—whom we know—and objects, which must be known. In ourselves, there are only four faculties that are of use for this purpose—*viz.*, the understanding, the imagination, the senses, and the memory. The understanding is the only one capable of perceiving truth; however, it needs the assistance of the imagination, senses, and memory so as not at random to dismiss anything that we may perceive.

As far as reality is concerned, it is sufficient to examine three things: that which spontaneously presents itself; how one gets to know one thing or object by means of another; and, finally, what deductions may be made in regard to each. This listing would appear complete, leaving out nought that may be covered by human effort.

A further thinker of the same era (midseventeenth century), Spinoza, maintained that God works by virtue only of the laws of His nature and is not forced by anyone. God understands an infinity of creatable things but will never be able to create them all, because if he were to do so His omnipotence would come to an end and He would become imperfect. Further, God's understanding, insofar as we conceive of it as the essence of the divine, is different from our own both in essence and existence; the twain, as it were, shall never meet except in having the same name.

If the movement perceived by the nerves of that seen by the eyes is somehow healthy, then the object or scene in question is known as beautiful; in the opposite case, ugly. Man judges things according to the condition of his brain and imagines them rather than understands them.

I understand under *idea*, a concept of the soul, formed by the soul because it is a thinking entity.

The human body is made up of many and varied individual parts, each of which is very composed.

The human soul is suitable for perceiving many things, the more so when there is more disposition on the part of the body.

The idea that constitutes the formal being of the soul is not simple but is composed rather of many other ideas. The idea of the human body is composed of all those numerous ideas of its component parts. The human soul doesn't know the body, nor is it aware of its existence except by the ideas of emotions that affect the body. The idea of the soul is united with the soul in the same way that the soul is united with the body.

The soul does not know itself except insofar as it perceives the ideas of the emotions of the body. The soul does have an adequate knowledge of the eternal and infinite essence of God.

In the soul, there is no absolute or free will; rather, it is determined to love this or that for a reason, which, in turn is determined by another, and so on to infinity. We function by the mandate of God, and we take part in the divine nature.

The philosopher and mathematician Leibnitz stated that just as the soul should not be used to account for the details of the economy of the body of an animal, so also such forms should not be employed to explain the particular problems of nature, although they may be necessary for establishing true general principles. The Monad is a simple substance, in the sense of having no component parts, and it forms or is a constituent of compounds. There must be simple substances since there are compounds, and a compound is but a group of aggregates of simple things.

It is, however, necessary for the Monads to have some qualities, otherwise they would not even be Beings.

It follows from the above that the Monad's natural changes come from an internal principle, because an external cause cannot influence its interior.

Moreover, it has to be admitted that Perception and that depending on it cannot be explained by means of mechanical parameters—that is, by figures and movements. If we were to conceive of a machine, the structure of which were to allow of thinking, feeling, and perceiving, it could be conceived of as bigger, such that it could be entered as can a flour mill. Once inside there would only be parts, acting one upon the other, but never anything that would explain a perception. So it is, then, that the seat of Perception must be sought in the simple substance and not in the compound whole, or machine. Thus in the simple that is all that would be found—that is, perceptions and their changes. Further, this is all that the internal actions of simple substances consist of.

It is also seen that the perceptions of our senses, even when clear, must of necessity contain something of confusion since, as all bodies in the universe are in sympathy with one another, ours receives an impression of all the rest; and while our senses address themselves to everything, it is not possible for our soul to become apprised of everything in particular. Accordingly, our confused feelings are the result of an infinite variety of perceptions.

For this reason, it can be seen that each living body has a dominant relative perfection which is the soul of the animal, but the members of the body are full of other living things—plants and animals—each of which, in turn, has a dominant relative perfection, or soul, of its own. There is no need, however, to consider that each soul has its own mass or portion of matter, assigned to it for all time, and consequently that it has other lower beings intended always for its service because all bodies are in a continual state of flux, as are the rivers, and parts enter and leave all the time. Philosophers have always been pressing for the origin of forms, relative perfections, or souls, but today, since it has been discovered by means of exact measurements on plants, insects, and animals, that bodies in nature are never the product of chaos or putrefaction but always of seeds of some kind wherein there is undoubtedly some sort of preformation, it has been considered that not only was the organic animal there before conception but also the soul—in a word, the animal itself.

The psychologist Freud declared that the richest sources of interior excitation are the instincts, which are representatives of all actions of energy coming from the interior of the body and transferred to the psyche; instincts, in fact, constitute the most important, and obscure, element of psychological research.

It would thus fall to the upper layers of the animus to bind together and relate the instinctual excitation characteristic of the primary processes. Failure in this mission would give rise to a perturbation analogous to traumatic neuroses. Only when the mission of relating together such excitations has been properly accomplished can the reign of the principle of pleasure or its modification, the principle of reality, be imposed. An instinct would be, then, a tendency, proper to living organic matter, toward the reconstruction of some previous state which the animus had been constrained to abandon under the influence of outside perturbing forces. This would be a sort of organic elasticity or manifestation of inertia in organic life.

The objection that in addition to the conservation instincts—forcing repetition—there are others that impel the being onward and upward, so to speak, merits serious consideration.

At some undetermined time, there were awakened, in inanimate material, by some unknown force, the qualities of life. This process may have been the one that served as a model for the later process that brought conscience into being in certain states of animate material.

To date, no one has been able to show a general instinct of super-evolution in the animal and vegetable world, although it would appear that such a tendency toward evolution is more or less established.

For many of us, it is difficult to dispense with the belief that in man himself there is an instinct for improvement which has brought him to his current high spiritual and ethical level and which can be expected to attend to his development, on up to the level of a superman.

More recently, Bertrand Russel's position was that what each man knows depends significantly on his individual existence: He knows that which he has seen, heard, and read, and what has been said to him and also what he has been able to infer from that data.

Theologists consider God as contemplating space and time from the outside, impartially, having a uniform knowledge of reality overall. Science endeavors to imitate this impartiality, with, apparently, some measure of success, but this success is partly illusory. People differ from the God of the theologists insofar as they have a *here* and *now*. The here and now is lived; the remote presents an obscurity that grows gradually. An atom only manifests its presence when it emits energy. There is a prudent silence in regard to where the atom gets its energy from.

103

What may happen to an individual atom in specific circumstances is uncertain. This is not only because our knowledge is limited but also because there are no physical laws which offer a specific result.

The theory of relativity—confirmed by experimental observation—shows that mass is not constant, as had been thought, but that it increases with velocity, and if a particle could move at the velocity of light, then its mass would become infinite.

As for human knowledge, we can ask two questions: first, What do we know? and second, How did we come by this knowledge?

To the first of these questions, science can respond, trying to be as impersonal and dehumanized as possible. In the resultant panorama of the Universe, it is natural to begin with astronomy and physics, which address themselves to the vast and the universal. On the face of it, life and the mind, strange in themselves, have little influence in the course of events and have to take something of a secondary position in this impartial examination.

In regard to the second question, however—how we came by our knowledge—psychology is the most important of the sciences. It is not only needed to subject the processes whereby we extract inferences from a psychological standpoint, but also the fact that all the data upon which we have to base our inferences are also psychological, inasfar as they are the experiences of isolated individuals. The apparently public, open nature of our world is partly illusory and partly inference. All the raw material of our knowledge consists of mental events in the life of separate individuals. Thus, in this case, psychology reigns supreme.

That which would commonly be called the mental life of an individual is entirely made up of ideas and attitudes toward them. Imagination, memory, desire, thought, and belief presuppose ideas, and the ideas are related to suspended reactions. Ideas are, in fact, parts of the causes of actions and are converted fully into causes upon the application of an adequate stimulus.

In practical science, there are two types of inferences: those that are purely mathematical and those that may be called substantial. The inference of the laws of Kepler from the laws of gravitation applied to the planets is mathematical; the inference of the apparent recorded movements of the planets from Kelper's laws is substantial, since the motion of the planets is the only hypothesis compatible with the events observed.

We also have the other argument regarding the brain and the mind: When a doctor examines a brain, he doesn't see thoughts; the brain is one thing, and the thinking mind is another altogether. The fallacy in this approach is in supposing that man can see matter. Not even the ablest physiologist can do so; his perception when looking at a brain is an event in his mind having but a causal relationship with the brain he imagines he is looking at. (I am using the word *thought* as a generic term to signify mental events.)

When we consider physical events in space-time, where there are no brains, we don't have a positive argument to show that they are not thoughts, except arguments arising from the observation of differences between living and inert matter, together with inferences based on some analogy or its absence.

Energy, although abstract, is a generalization arrived at through quite concrete experiments, such as those of Joule. Physics, as a verifiable discipline, then, uses empirical concepts in addition to the purely abstract concepts necessary in pure physics.

All nonmathematical terms used in physics—considered, as it is, an experimental science—have their origin in our sense experience; it is only because of this that sense experience can confirm or refute physical laws.

The practical use of science is in its capacity for predicting the future.

It is now possible to conceive of the final structure of the physical world not as a continuous flow, as in conventional hydrodynamics, but more in a Pythagorean fashion, whereby models arise from analogies bit by bit. Evolution—which in Darwin's time expanded, as a concept, little by little, from precedent to precedent—now progresses in revolutionary leaps, with mutants and/or abnormal beings. Maybe, with our wars and revolutions, we have become impatient with gradual matters. For whatever reason, modern scientific theory admits the brusque and abrupt more readily than did the Victorians, with their imagination of a softly flowing cosmic current of ordered progress.

As man has progressed in the application of his intelligence, his habits of inference have been more and more in keeping with the laws of nature. This has made his approach more frequently a source of true rather than false expectations. The forming of habits of inference, which give rise to expectations fulfilled, is part of the adaption to the environment, upon which depends biological survival.

PHYSICAL ANTECEDENTS

Here we are talking about the most interesting antecedents, in the light of quantum mechanics. First of all, we might mention that quantum mechanics concentrates mostly on explaining and formalizing effects; causes remain unknown.

If the gravitational and electromagnetic fields are unified through, as a minimum, an acceptance of the hypothesis of the similarity of phenomena in the U and e domains, should we then even suppose that in the e domain there would be physical laws different from those of the U domain? If there is a way to explain the causes, what should be done?

The first thing is to explain phenomena in the light of the new criteria and then to continue with the experiments and theoretical developments applicable to the e level and, as far as possible, carry on into the ee domain.

In this way the available mathematical apparatus can be cleaned up and simplified and quantum mechanics rendered more manageable, thus allowing it greater development.

This is no simple matter, and such updating may take years. Let us look at the present situation.

In classical mechanics the dynamic state of a system at any moment in time is determined via a knowledge of the position and movement of each of its particles. In quantum mechanics, the simultaneous knowledge of these coordinates is not possible, and it can only be hoped that both types of description coincide at the deterministic limit of quantum mechanics.

In quantum mechanics, description of the state of a system is fully described by its wave function. One can try to recuperate the classical picture by attributing to each particle position and movement coordinates that are precisely the mean values of the corresponding observable ones in the state under consideration and dismissing the fluctuations about these values. Under certain conditions this can be done.

Further, in order that the movement of a wave packet may appear similar to that of a classical particle, it is necessary, as has been said, that two conditions be fulfilled: The means of the observable position and moment values must satisfy, to a good approximation, the classical laws, and the dimensions of the wave packet must be sufficiently small and remain so during its development.

In general, however, the study and understanding of phenomena in the e and ee domains in the light of quantum mechanics is not simple. To see this, we have but to look at its first postulate: To each physical system to be described within the framework of quantum mechanics, there must be made a corresponding, complex and separable, Hilbert space. A pure state of this physical system at a moment t is represented by a unit beam belonging to the corresponding Hilbert space. An element of the beam is called a *vector state*, or *ket*.

It is apparent that it would be beyond the frame of reference of this essay to go further on the matter. For instance, in order to simplify the formulae that appear in the calculation of atomic structures, it is useful to use Hartree's system of atomic units (a.u.), whereby the units are established via those of the electron and the action quantum. As for the atomic nucleus, given the small mass difference between the proton and the neutron (1.4 x 10^{-3}), these particles are, from the U domain, indistinguishable and are referred to under the general name of *nucleon*.

One of the basic problems in trying to throw some light on phenomena of the e and ee domain is that of the study and calculation of the transmission and reflection coefficients, of a particle of energy E, as caused by a specific potential barrier. For this, the old Sommerfeld-Wilson-Ishiwara quantification method may be used.

In quantum mechanics, we can not yet adequately define trajectories. Thus, if we consider a system of identical particles, so as to simplify matters at any particular instant, we can take the wave function representing its state as being given by the product of individual wave packets. These generally overlap, so that on locating a particle at a later instant, it is quite impossible to say which of the original particles it is. It is this wave nature that gives rise to the fundamental difference between deterministic classical mechanics and the indeterministic quantum variety as far as "discernability" is concerned.

If, for example, we wish to study the passage of light through a crystalline lattice in refraction or light-decomposition phenomena, we must use the atomic vibration exitation quanta of the lattice—called *phonons*— for the interaction with photons. In general, to describe the interaction of light with matter, the first in the e or ee domain and the second as we see it from the U level, covariant equations are used. These are the same as those that allow of an explanation of the radioactive transitions in atomic and nuclear systems in the case of this approach to development of the quantum theory of radiation.

There is a similarity in the area of the absorption and emission of energy and the appearance of photons: There is spontaneous emission of a system when it drops to a lower level of excitation. Originally there are no photons at all; then one appears. In the same way, there may be an initial state of electromagntic radiation and finally, apart from the emission, an absorption at the expense of the initial electromagnetic energy.

There is the case in which this is no energy transfer from the photon to the electron. This is observed in the so-called Compton collision at low temperatures, concerning the elastic collision of soft photons with free electrons practically in repose.

Most of quantum mechanics is devoted to the study of particles in electromagnetic fields—that is, to interactions between domains. It leaves little margin for a separate study of phenomena in each domain, which would mean that we simple mortals, in fact, for a series of practical reasons, started out along the most difficult path. Even so, if we bear in mind that almost all the information we have about the phenomena in question was obtained through just these studies and experiments, then maybe the historical development was justified.

The problem of measuring parameters of the microcosmos is a very real one; measurements are always of effects and almost always in some indirect fashion.

The central problem of measuring lies in the reconciliation between the physical states of a system according to whether we are concerned with a measuring process or an evolution or development over a series of consecutive measurements. In the future, to the extent that feasible laboratory experiments can be thought up—such that they give direct measurements of the causes of phenomena of the microcosmos—there will be significant advances.

I believe that if the fields are unified, then step by step, all of quantum mechanics may be revised such that the theory will be unambiguous in the matters of evolutive and collapse phenomena in the e and ee domains. This would thus give this theory—so fundamental for the progress of physics and, in the final analysis, for all humanity—the status of a deterministic science of the same kind as classical mechanics.

A priori, we know it would be more complex, since we are trying to establish from a higher domain, U, laws for phenomena going on in the lower domain, e, and this via methods of measurement and how we, from the U domain, interpret them.

In order to have some mental setup—practical, if possible—of the number of unknown variables, the so-called hidden variables were invented, thus to fill in the gaps in our knowledge, not to speak our ignorance. There is, however, von Neuman's theorem, according to which there are no hidden variables. This is that the stochastic aspect of quantum mechanics cannot be reproduced on the hypothesis that there are parameters, analogous with the positions and velocities of the molecules of a gas, the knowledge of which will fix the microstate of a system, free of dispersion for all observables and in terms of which a pure ordinary quantum state would not be more than a macrostate, like an incoherent mix of such parameters. Later, though, the idea of accepting the results of this theorem *a priori* was considered unacceptable because it was a way of trying to show an impossibility and hence closing that way of development of knowledge. The hidden variables were to be kept as variables of our ignorance, establishing, in turn, which conditions we should not demand of microstates. For this reason, the hidden variables have to circumnavigate the stochastic predictions or principles of local action.

The local-action principle of determinism cannot be accepted if such determinism has not been achieved—that is, for lack of experimental data. While we are in a transition state between indeterminism and determinism, the local-action principle that does not fall down in the area of what is observable today, even with corrections and relativistic principles, may well do so for the hidden variables belonging to an area that, while relativistic principles are applied, is that of a different domain. In this domain, the local-action principle is affected by such enormous domain-passage constants that even imagining a phenomenon to try and give it the correct interpretation would require a great effort.

Chapter XI

INTELLECTUAL CAPACITY

We should underline the limitations of the intellectual capacity of man, which is made manifest by the unimaginable (infinity, God, *et cetera*,) and we have to look at ways of coping with this situation. Otherwise we would have to have a premise of the limited capacity of man to overcome some limitations to knowledge, which, of necessity, he is going to confront.

There is a tendency to believe in some beings, superior to ourselves, living on other planets. There are even tendencies beyond that of supposing that they may be the gods in which men have always believed. If this were the case—that they were the gods of the U domain—then what goes on in UU?

We feel that God, the supreme fount of answers to our "final" questions, is to be located in UU, and as our knowledge advances—when we find out what goes on in UU—then God will be located beyond that again.

Today, the location of the gods on Mount Olympus (and many other places) is still fresh in human memory. Later came the sole God in heaven, and as we go further afield God is always located just beyond our horizons.

If we see our limitations so, then we have to see what is the optimum use to which they may be put and any way there may be of improving them.

We are not talking of an optimum in the sense of an adequate social use, or the use to the best advantage of the brain.

We know there is a large part of the brain having unused capacity and that, with some cultural means and ways, man may be able to use that capacity. This would also signify a limit to the intellectual capacity of man. But what if the problems to be faced in the future are beyond this limited capacity? What chance does man have of overcoming himself?

I have no answer, apart from that in Chapter VI—namely, that man may have to rely on computers. This method of eliminating tedious work and using computers, which are very good at it is, however, is hardly a solution and less a good solution. In fact, it is a bad one, but you have to start somewhere.

It is virtually certain that in the future there will be programs for almost everything, but how can a machine have imagination such as that of men?

I can only see the method mentioned whereby we get surprising answers—answers generated by the imagination of the computer. By this computer "imagination" I mean that, given the constituting elements available, the machine generates all the possibilities and gives us both the imaginable and the unimaginable ones.

I want to mention the subject of imagination and instinct in the light of the supposed similarity between the domains. In this, I want to bring in the similarity we established for living matter. Thus, if those beings have results like those we have at the U level, then imagination would be the messages they send us of their impressions in the framework of the data they have from their own domain and data received from our domain. Instinct would probably be the same—messages at the conscious level sent when the beings considered it appropriate as a function of data they receive from the U domain. For example, we instinctly withdraw our hand from the area of a fire which never touched us, because our eyes transmitted information of the danger. This is different from the automatic reaction we experience when the fire does touch us.

Chapter XII

SUMMARY

The various areas close to the frontiers of physics touched on here, and what we can glimpse ahead, have taken us on journeys through the micro-cosmos, the universe, and beyond.

In the way they are stated, they are, for me, an interesting guide for research since, if the suppositions have some basis, it must be determined whether they serve to explain all the phenomena we know and if they may form the basis for an advance in physics or not.

Summarizing, I believe I have put forward the following general points. Thus I think that

(1) I have adopted an adequate reference system with the domains, similar to what the introduction of systems of reference coordinates represented in their time (Chapter I);

(2) I have made an improvement on gravitational theory and opened a way for the use of antigravity (Chapter II);

(3) I have set down a basis for breaking the impasse in the problem of unifying the fields (Chapter III);

(4) I have given more solidity to the wave-particle duality (Chapter IV);

(5) I have made a new proposal, on a physical basis, in regard to organic and living matter (chapters VII and VIII);

(6) I have made mention of the limitations of the brain of man and suggested a way in which these may be overcome (chapters VI and XI).

While history teaches us that one should expect no special recognition for introducing new concepts, within the complexity of the subject matter, I do hope to have made some small contribution.

What would it be? I really don't know—the gravitational formula and its consequences, the unification of the fields, or maybe a way for theorizing on the physical basis of life. Perhaps one of these will serve as a guide when, at present, there is already genetic engineering in the complexities of which the initiated move as though in a dark tunnel without a key to the code that may reveal the secret lying at the end?

There may be something in the consistent quantification of biology at the atomic level or also in the God necessary for millenia to come for the consolation of so many and the fight against merciless atheism based on the ignorance it claims to struggle against.

Is the suggestion for a physical basis of psychology of use? It gives a location and an identification of the soul in our own microcosmos and brings us back to the historical admonition to know thyself!

Recapitulating, we followed this route: We showed that Newton's expression is a particular case of a more general situation—action at a distance is not action but an effect. Once we accept a similarity between phenomena in the cosmos and the microcosmos, we can see electromagnetic phenomena as comparatively high-energy gravitational happenings of the microcosmos.

We show the gravity constant as the passage value of the resultant of the general energetic gravitational state—namely, that of the energy of which the masses, of constant value, are the carriers—because gravity although it can form systems in equilibrium, cannot cancel itself out just because it is a manifestation of energy.

We mention that if there is such a constant for the cosmos, there must be another for the microcosmos—namely, the value of the resultant of the general energetic gravitational state for the subatomic level—and that the product of the two forms a total universal gravity constant up to the one we have.

We also present the use of such constants, according to the phenomenon concerned, as giving interchangeable expressions for phenomena— that is, as though everything were a unified field of phenomena and physics were a single whole.

We apply the above to some well-known formulae, starting to unify fields that to date have been incompatible. We broadened these concepts to take in organic and living matter, so as to explain their known manifestations.

As for the quantification of biology, matters are more difficult, but the basic principle that clarifies the panorama is put forward.

All this should be continued. In physics, a body of formulae in the unified field has to be generated, and from a study of such expressions there will arise a clearer and more definitive understanding of physical phenomena. This can only give rise to scientific and technological advance.

I want to say one last word about the use man makes of scientific discoveries. In times past, primitive man, on lengthening his arm with a club, used it to kill both food and his neighbor. The "scientist" who invented the club did so for the first application, while man used it for the second.

Today, when science provides us with more sophisticated clubs, it should not mean that we should cease to invent them. It is we who use them wrongly, thereby betraying the principles of humanity.

So it is that now, as then, the only solution is to be vigilant in the fulfilment of moral principles by man, looking after others. Such vigilants must come from all walks of life—patrician and plebeian, rich and poor, and proletariat and the capitalist, peacemakers and warriors, anarchists and democrats—in fact, every man and woman on the planet; only the insane are excepted.

Part Two

Matter
and Movement

This essay does not allow any present physics teaching program, so some comments are necessary in regard to the use that may be made of it.

It is the continuation of my previous essay "The Unified Field" and is, using the principles therein suggested, a revised general physics course. The underlying purpose has been to interpret phenomena assuming, in physics, a unified field which, we arrive at from

(1) a quantification of domain passage,
(2) The equivalence of Newton's and Coulomb's equations,
(3) The use of the author's K and K_p constants, and
(4) The postulate, in masses, of the distribution of velocities throughout the domains.

In the unified field, several expressions may be used for a single phenomenon, so I have chosen the ones that seem most promising without wishing to say that the approach given here is the only one useful for any particular case. Besides, since this is a personal interpretation of matters, I had the option of choosing which expressions would be given priority.

Furthermore, since the thrust here was to come up with a conceptual basis of phenomena, I have avoided, as far as practicable, concrete calculations, limiting myself only to those that were indispensable for illustrative purposes.

The chapter sequence was set up so as to give a certain continuity of the reasoning used. Thus it is assumed, when starting with mechanics, that the reader has some idea of astronomy and cosmogeny. In the following chapters, it is assumed that the reader is more or less familiar with the corresponding areas of general physics—first, to grasp the subject matter, and second, to have an idea of the benefits that may flow from better knowledge of the phenomena I hope to be presenting.

I will be referring to the essay "The Unified Field" and assuming the reader is familiar with it; when necessary, however, a short summary of the pertinent points will be given.

I have classified all physics according to the phenomena occurring in potential fields, this being the best way to set up a systematic study in line with the criteria listed above.

While the introduction of the constant K allows of a more detailed study of phenomena, the essay has not been titled "The Unified Field" since unification of itself is a dynamic process, and the intention here was to emphasize the role played by matter and movement.

In understanding and subsequently quantifying the very peculiar structure of matter, we will begin to divest ourselves of our prejudice in thinking of matter as being everything and movement nothing more than a relationship between a couple of dimensions. Indeed, we will find that movement is everything and matter almost nothing—virtually an illusion (albeit tangible).

This is nothing incredible—only a little awkward—to understand. Once the leap is made, however, one can only marvel at the astuteness of the Creator in the structure of matter: We find a perfect machine—a *moto perpetuo*; there are simply no losses, either of matter or movement. This would, of course, be in conflict with the conservation principles. If mass is to be set at nought and movement is produced by a force, then everything is reduced to forces.

What, then, is force if not mass by acceleration? If we are to dismiss the mass for the reasons stated and remember that acceleration is a variation of velocity, then we find that force is based on the same thing it has been used to explain, namely, movement.

We then run again into one of the Creator's little traps. For three centuries the velocity factor has been hidden in Newton's equation. Now that we want to use the equation to study the successive structure of matter, we find a new problem: If everything boils down to forces that produce movement in successive domains, then the "ultimate" force has to be that which is to be found in the last and lowest domain.

If we make an assumption in an effort to locate some answer, we don't necessarily have to go into detail to explain what it is about. We don't know—it may be just another twist; in fact, I think it is. If God doesn't play dice, it's because such a game would be below Him; His game is more sophisticated, but what a game! And we are figuring it out!

The reader will hardly fail to notice that, with the explanation of gravity given here, this work has of necessity to revise all the bases of physics. This is the type of risk that can give rise to great success, or proportionate failure.

What do we mean by the *bases of physics*? Here we are referring to those of mechanics, since they form the basis of the rest.

If we recollect that the bases of mechanics are laid down in Newton's *Principia*—not to mention the work of Galileo, Huygens, Einstein, et al—are we here trying to rewrite the *Principia*?

That would be quite a project, and I shiver at the prospect. Accordingly, I have called this an "essay." Even so, if we can no longer accept the concept of attraction in the traditional sense, then, to be consistent, we do have to revise its bases, in the same way Einstein did for absolute time when, with inspiration and perspiration, he showed it to be an untenable concept.

With this essay I believe I have laid a foundation for the development of a new physics. In "The Unified Field" I have a simple, popular, and—possible—somewhat emotional presentation to explain the changes that may soon come about. In this second state, I have tried to show how the exterior aspect of such a physics would appear, giving some comments on its significance and implications.

Chapter XIII

THE POTENTIAL FIELD OF SOLID BODIES IN THE U DOMAIN

DEFINITION

We shall define the domain of the whole universe as U. By the whole universe we mean the entirety of the region within which all the laws we employ are valid; such laws would be those of physics, biology, and suchlike, along with Newton's law of universal attraction and the geometries, both Euclidean and non-Euclidean—in sum, all those laws that do not predominate at the atomic, molecular, or ionic level. In this domain, which we shall define as e, the laws of electromagnetism, thermodynamics, and quantum mechanics and suchlike apply.

The subatomic domain, in which, because of our ignorance of the applicable laws, we have no such laws to apply, we shall define as ee.

If we move to the other extreme—namely, beyond the limits of the universe U as it is known to cosmogeny—then we need a further designation, and we shall employ UU.

While for the domains UU and ee, we have no principles or laws of phenomena insofar as we are unaware of what may take place there, I want to take some inadequately explained phenomena in the domain U and e with a view to inferring what may be going on in ee. In the same fashion and as a function of what is happening in ee, e, and U, we should be able to begin to have some idea of what happens in UU.

Thus each domain is contained in another that precedes, and contains another one following it, just like a Russian doll. We could even refer to this as the Russian Doll Syndrome.

CHANGE OF SCALE

For passing from one domain to another, the domain-passage constants in the case of all parameters are those resulting from the quotient of each particular parameter in a domain with respect to the correlative parameter of the corresponding preceding or following domain.

The domain-passage constant with a value of unity is that for velocity, arrived at by considering the principles of conservation of energy and distribution of velocities in masses.

The change of scale must in each case be determined, and to go into examples is beyond the purpose of this essay insofar as the procedure is somewhat extensive.

In the process, however, the equations of relativistic physics must be taken, and simplified if necessary. The inverse procedure should not be applied.

$$\frac{\text{Distance in U} \div \text{Distance in e}}{\text{Time in U} \div \text{Time in e}}$$

$$= \frac{\text{U/e distance scale}}{\text{U/e time scale}} = \text{Velocity scale} = 1$$

Consequently, the Distance Scale is equal to the Time Scale, and if one were to vary, so should the other.

If the velocities in a domain are such that the relativistic correction with respect to the reference systems considered gives a distance variation, then on passing to another domain the time scale will vary with respect to the time scale which would have corresponded to another velocity. The same applies in reverse—if there is a variation in time, the distance scale will change. This means we are dealing with a dynamic system and a different passage scale applies for phenomena being compared in different situations, except for the case of velocity.

GRAVITATION

Definition of a potential field: This is given by the energy content of a body by reason of its position with respect to another body.

We accept that the force of gravitation between two bodies is given by the following expressions.

$$F = \pm\, G\, \frac{M\overline{V}m\overline{v}}{d^2} = \pm\, G\, \frac{Mm\, \dfrac{\overline{V}}{\overline{v}}\, \dfrac{\overline{v}}{\overline{v}}}{d^2} = \pm\, G\, \frac{Mm\, \dfrac{\overline{V}}{\overline{v}}}{d^2}$$

$$\boxed{F = \pm\, G\, \frac{M\, m\, \overline{V}}{d^2\, \overline{v}}} \qquad \text{(XIII-1)}$$

$$\pm \int dF = \frac{G.M.m}{d^2} \int d\,\frac{\overline{V}}{\overline{v}}$$

121

$\overline{V}$ = the sum of all the velocities of the greater mass, as measured in U
$\overline{v}$ = the same for the lesser mass
We also accept

(1) when $\overline{V}/\overline{v}$ = 1, there is a constant separation in space between the bodies;
(2) when $\overline{V}/\overline{v} < 1$, this separation decreases with time, producing an approach (or presumed attraction); and
(3) when $\overline{V}/\overline{v} > 1$, this separation increases with time, (giving a presumed repulsion).

If we also accept that the gravitational phenomenon in the U domain is quantified by G and that G = f(velocities), then we can say that the gravitational force is a function of momentum and the gravitational phenomenon is a function of velocities.

We find a precursor of this in Mach's experiment of the two reacting cars, in which $(m_1/m_2) = (V_2/V_1)$, or $m_1 = m_2 (V_2/V_1)$. This is to say that masses are related to the quotient of their velocities.

The law of gravitation could be stated as follows: "Each particle in a domain belongs to the potential field, which determines the momentum it has, and is related to the potential fields of the other particles of the domain by a force proportional to their respective momenta and inversely proportional to the square of the distance separating them."

If the gravitational phenomenon is a function of velocities, we can write

$$\boxed{G = f(v)} \qquad . \qquad\qquad \text{(XIII-2)}$$

A corollary of the above is that the state of weightlessness is a function of velocities, independent of masses, and is produced when the ratio $\overline{V}/\overline{v}$ is equal to unity.

The basis of this may be found in today's Cosmogeny, in which it is accepted that:

At the beginning of time, there was a certain amount of mass in the universe, carrying a certain amount of energy in the form of velocity. The mass was dispersed into many parts, the result of a vast explosion, which continues and within which we now live. Although the universe is exploding, we do not normally see it as such because of the change in time scales; it happens before our eyes as though in very slow motion. Each portion of mass, then, has its own velocity.

When we look at the structure of the mass, we find it strange insofar as it is almost empty; furthermore, a closer look reveals a spatial structure within this emptiness, similar to that found outside of it. If this appears complex but acceptable, with some effort and in the light of experimental evidence, what should our reaction be when, on looking even closer, we find in the inner particles a further repetition of the same structure? Having overcome our surprise at this apparently ongoing phenomenon, we readjust our understanding and draw the conclusions that follow.

Before the Big Bang, this inner, and very particular spatial structure was under tension because of the enormous concentration of mass. When the pressure build-up of agglomeration came to an end, the whole exploded rather in the fashion of a spring on release from tension. In the innermost parts there was further tension, and in the explosion we observe this tension (energy) passing from inner parts to outer parts.

It should be realized that all energy-transfer phenomena are of this type—namely, from inner parts to outer parts. This would include solar energy passed to the earth, radioactive transformations, animal metabolism, and even clouds moving in the sky. Anything that we observe forms part of this process, which, fundamentally, is the transfer of energy (velocities) from inner to outer parts.

The overall tendency is toward a possibly uniform distribution of the amount of available velocity in the totality of the mass available in the universe. When the process is complete we do not know what kind of situation will prevail. The fact is, we are also uncertain as to whether, in the process, there is any addition to or subtraction from the mass or velocity of the sum, both of which may take part in the event. There is also doubt regarding the initial Big Bang itself; it is, after all, but a postulate, which is acceptable for deducing the rest and explaining what we observe today.

Now, what happens with masses that give up velocity—where do such masses come to rest? Let us consider the example of a dam. If we have water in a dam 1000 meters high, it has a potential energy of X. On release of the water, in a controlled manner, through the turbines located below, its mass gives up energy and thereafter the water issues from the turbines and flows peacefully off down the riverbed at the base of the dam.

What happens in space—what constitutes the riverbed for a mass that gives up velocity? Where does such a mass encounter a lower point and equilibrium with the level of potential proper to itself? The answer would appear to be straightforward: we have already seen that in "near space" there is just such a potential field of equilibrium. An example would be the moon losing velocity and falling toward the earth; if the earth had a hole large enough to receive it the moon would come to rest at the "riverbed" of the center of the earth. The process could be repeated with the earth coming to rest at the center of the sun, which could come to rest at the center of the galaxy, and so on.

Could such a situation give rise to a vast agglomeration followed by a further Big Bang? We don't know; all we know is that velocities tend to distribute themselves. Thus bodies are not attracted by the earth but are subject to the tendency to distribute velocity and lose energy and so seek a lower point of rest.

Why does an object fall toward the earth? First, because it does not have sufficient velocity to remain at the level in which it is located; and second, in view of its lack of sufficient velocity, it seeks the energy "plane" proper to it, unless prevented from so doing by some obstacle. The obstacle itself, though, is giving energy to the object by retaining it at some particular level. This again is a process that can be carried through with the obstacle of the obstacle, et cetera, at each stage describing the mechanics of equilibrium.

To follow through with this, it is necessary to go from domain to domain and, in the observable universe, locate phenomena in the scheme of events. This is given to us by the laws of the tendency that those phenomena obey.

WHAT IS MASS?

As we consider it in the U domain, mass consists of body and substance, and in the solar system it occupies 2.4×10^{-25} percent of the space. Substance made up of atoms, however, probably occupies at the e level, without going on to the ee level, a similar amount of space. We can see then that taking the two domains, the substance occupies some 2.4×10^{-2} percent of 2.4×10^{-25} percent—or 5.8×10^{-50} percent of the available space.

If we were to take more domains into account, this percentage would drop even further. Indeed, if we take one or two domains more we finish up with almost a total vacuum, and mass would virtually not exist; for a very small m, we have

$$\boxed{F \sim m\,v \sim v} \quad \text{and} \quad \boxed{F \leftrightarrows v} \,. \qquad \text{(XIII-3)}$$

The double arrow represents a conversion, or virtual equality.

Even so, this mass would contain all the energy by virtue of the velocities of its particles.

The expression mv is the simplest expression we have for expressing energy, insofar as it is a force by a distance and a time unit.

In this expression, if we assign the mass the function of energy carrier it becomes an invariant.

Now, if an electron accelerated has more energy than before, this is a result of the new velocity (mv, or, as happens in phenomena in which the ee domain plays a part, mv^2) not an increase of mass.

The expression mv^2—or the Vis Viva of Huygens, similar to Einstein's $m\,c^2$—may be explained as the mass having a velocity V, of the e domain, and another velocity, V′, of the ee domain, with the energy E measured from U. This gives rise, in ee, to the mass formed by e-level subelements having a certain V′, which is applied to the new mass in the e domain formed by them and which has its own V in the e domain; these are multiplied because each ee-domain V′ unit is applied to each V unit in e.

The Russian Doll Syndrome should help to grasp the phenomena.

$$V\,V' = V^2 = c^2 \qquad\qquad \text{(XIII-4)}$$

In this situation, we have $E = m(KG)/h = G'/h$, and remember, as per (XIII-2), that $G = f(v)$ and that G'/h is the amount of action (energy) of the phenomena to be manifested in U, we have $E = m\,v^2 = m\,c^2 = m(KG)/h$; hence $c^2 = KG/h$. Thus equation (XIII-1) may be written as follows:

$$F = \pm\,\frac{E_1\,E_2\,h^2\,\overline{V}}{G\,K^2\,d^2\,\overline{v}}\,. \qquad\qquad \text{(XIII-5)}$$

Other forms of expressing F would be the following.

$$F = G\,\frac{Mm\,\overline{V}/\overline{v}}{d^2} \qquad\qquad \text{(XIII-6)}$$

$$F = \frac{h\,c^2}{K}\,\frac{Mm\,\overline{V}}{d^2\,\overline{v}} \qquad\qquad \text{(XIII-7)}$$

$$F = \frac{h}{K\,\mu_o\,\epsilon_o}\,\frac{Mm\,\overline{V}}{d^2\,\overline{v}} \qquad\qquad \text{(XIII-8)}$$

$$F = G\,\frac{\dfrac{E_1 h}{G'}\,\dfrac{E_2 h}{G'}\,\overline{V}}{d^2\,\overline{v}} \qquad\qquad \text{(XIII-9)}$$

$$F = G \frac{\left(\dfrac{G' \, \mu_o \, \epsilon_o}{\lambda \, v}\right)_1 \left(\dfrac{G' \, \mu_o \, \epsilon_o}{\lambda \, v}\right)_2 \overline{V}}{d^2 \, \overline{v}} \qquad \text{(XIII-10)}$$

$$F = \frac{h \, c^2}{K} \frac{\dfrac{E_1 h}{G'} \dfrac{E_2 h}{G'} \overline{V}}{d^2 \, \overline{v}} \qquad \text{(XIII-11)}$$

$$F = \frac{h \, c^2}{K} \frac{\left(\dfrac{G' \, \mu_o \, \epsilon_o}{\lambda \, v}\right)_1 \left(\dfrac{G' \, \mu_o \, \epsilon_o}{\lambda \, v}\right)_2 \overline{V}}{d^2 \, \overline{v}} \qquad \text{(XIII-12)}$$

$$F = \frac{h}{K \, \mu_o \, \epsilon_o} \frac{\dfrac{E_1 h}{G'} \dfrac{E_2 h}{G'} \overline{V}}{d^2 \, \overline{v}} \qquad \text{(XIII-13)}$$

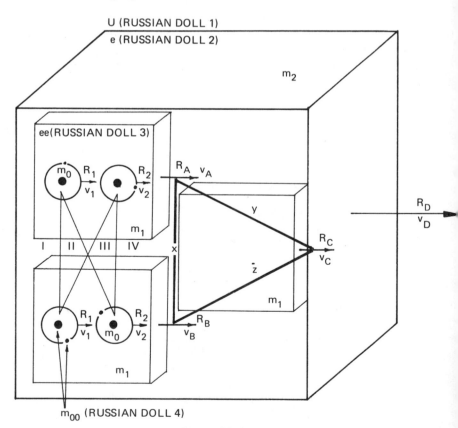

Figure 36

126

$$F = \frac{h}{K\,\mu_o\,\epsilon_o} \; \frac{\left(\dfrac{G'\,\mu_o\,\epsilon_o}{\lambda\,v}\right)_1 \left(\dfrac{G'\,\mu_o\,\epsilon_o}{\lambda\,v}\right)_2^{\overline{V}}}{d^2\,\overline{v}} \qquad \text{(XIII-14)}$$

The potential field 1 of ee has the resultants R_1 .... R_i and the same may be said of all the others, such that

$$\sum_{1}^{n} R_i = R_A \text{ and } R_A \equiv V_A$$

where I – II – III – IV are the reciprocal influences or applications.

The potential field of e has the resultants R_A, R_B ..... R_j and the same may be said of all the rest such that

$$\sum_{A}^{m} R_j = R_D \text{ and } R_D \equiv v_D$$

where X, Y, and Z are the reciprocal influences or applications. If this is the case, we would have

$$E_{eee} = m_{oo}\; 0 = 0$$
$$E_{ee} = m_o\; v_1^0 = m_o$$
$$E_e = m_1\; v_A^1 = m_1\; c^1 = m_1\; c$$
$$E_U = m_2\; v_A^1 \cdot v_D^1 = m_2\; v^2 = m_2\; c^2$$

and in general

$$\boxed{E_i = m_j\; v^j}. \qquad \text{(XIII-4')}$$

This yields the following conclusions:

(a) The eee domain is the apparent location of the elementary units of mass, m_{oo}, and force. We think of the mass there as corpuscles or blocks having no space inside them and force as mutual action between them, without, at present, being able to say anything further on the phenomenon.

(b) In the ee domain, we have the first mass, m_o, with unit velocity $V^o = 1$, whereby we obtain $E_{ee} = m$, which, from the U level, we express as $hE = mG'$.

(c) In the e domain, we have the second mass, m_1, which is multiplied by the unit velocity and which we must multiply by its own velocity (for example, $R_A \equiv N_A$), obtaining $E_e = m_1 v$. Now, if v_A is equal to zero, we obtain the m_o from the relation $E = (m - m_o) c^2$, where m is the same m_1, having a certain velocity acquired from a potential field outside the system.

(d) In the U domain, we have the third mass, m_2, which is multiplied by the unit velocity, by that of the e domain, and which we must multiply by its own velocity (for example, $R_D \equiv v_D$), obtaining $E_U = m_2 v^2$.

In e and ee, the own velocity values are c. We then use $v = c$, on the basis of equality of distance and time scales.

From UU, the E of the U domain would be expressed as $E = mc^3$. Try showing it.

POSTULATING PRINCIPLES

First: The mass principle: "Mass has value but barely exists. The total of all mass values is a universal constant."

Second: Principle of conservation of forces (or their equivalents, momentum, or velocity amount): "The total amount of forces is a universal constant."

Third: Principle of the distribution of velocities or conservation of energy: "In domains not in the state of total repose, velocities tend to distribute themselves among the masses, from lower domains to higher ones, during expansion phases, and in the opposite direction during contraction."

Does the tendency toward the recessive process, or that of ceding velocities, indicate that these processes are simultaneous?

We have assumed the current phase to be one of pure expansion. Even so, if the entire universe does not function as a single whole, some regions may be expanding while others are contracting.

We have opted to call the above-mentioned principle that of velocities distribution, or conservation of energy, because when velocities are distributed among the domains, as indicated, then we find ourselves in agreement with what we normally refer to as the conservation of energy in the U domain.

The structure of mass and the velocity passage permit the conservation of energy.

128

THE LAWS OF MECHANICS

The first principle has no corollaries but the second one has—namely, Newton's Third Law, and the First Principle of Thermodynamics. "All actions have an equal and opposite reaction—that is, the actions between bodies, of one upon the other, are always equal and in opposite directions." Consequently, force is a mutual action between bodies, and this is as far as we can get.

In the same way that we saw how mass becomes diluted—almost to the point of a vacuum—so also force, which, in its essence, is confused with velocity, becomes diluted in the concept of mutual action.

In view of the above, we shall not be able to know either mass or "velocity-force" any better until we are able to determine the real structure and phenomena of the ee domain at least.

The third principle has the following three corollaries: Newton's First and Second Laws and the Second Principle of Thermodynamics.

Corollary Number 1; Newton's First Law: "All bodies remain in their state of rest or uniform rectilinear motion unless forced to change this state when acted upon by an impressed force."

These forces may be either inside or outside the body, although we have always seen this law as functioning with outside forces. We should, however, consider cases where forces originating in lower domains are involved. For example, if the emission of light exerts pressure in a manner other than uniformly radial (how?) would be subject to a thrust originating from an interior force.

Corollary Number 2; Newton's Second Law: "A change in movement is proportional to the force applied and is in the same direction as the force. The momentum is proportional to mass and velocity jointly."

$$\frac{d}{dt}(m\,v) \; \alpha \; F$$

Since, by the First Principle, mass is an invariant in the relativistic correction, then

$$m = \frac{m_o}{\sqrt{1 - \dfrac{v^2}{c^2}}} \quad ,$$

where m_o is the hypothetical mass in repose, since, because of the structure of mass, m_o is a mass that has a characteristic velocity in each separate domain.

In other words, recollecting the Russian Doll Syndrome, we can say:

m_o = mass in repose in ee, seen from e

m = e domain mass, seen from U

$m - m_o = m_{ee}$ (submass of e) (m_o in Figure 36).

This means that the difference of masses in the kinetic energy calculation is equivalent to a mass arising as a component, which it is, of the ee domain.

$$E_c = (m - m_o) c^2$$ (XIII-15)

Also, the conversion of mass into energy and vice-versa has relative meaning and is only valid as a quantifiable equivalent:

$$h E = m G,$$

since, apparently, energy would be a force and mass almost nothing, but those "almost nothings" in reciprocal action would give rise to force.

Let us see this more clearly: Expression (15) is applicable to the e domain, and the difference of two masses is another mass:

$$m - m_o = m_y$$

all of which may be defined in terms of the velocities carried:

$$m - m_o = v - V_o = \Delta v$$
$$m_y = \Delta v$$

and we may then more properly write

$$E_c = \left(\frac{E\,h}{G\,K} - \frac{E_o\,h}{G\,K} \right) . c^2$$ (XIII-16)

where E = $m\,c^2$ in the e domain seen from U, and
   $E_o = m_1\,V^1$ in the e domain.

Now, V^1 equal to zero represents the condition of repose for the e domain mass seen from U, which is defined as m_o, or

$$m_o \sim m_1\,V^1.$$

130

Thus we have the whole phenomenon measured from U.

If we suppose that light is an ee domain phenomenon, seen from U, then its energy should be given by

$$E_{ee} = m_o \, v_1^0 = m_o \, 1, \qquad\qquad (XIII\text{-}4'')$$

whereby the condition $v_1^0 = c$ should be fulfilled. However, since v_1^0 was the unit velocity of passage between domains, it turns out that light has the "unit velocity," according to how the mass is constructed:

$$E = \nu h = mv = mc \quad \text{since}$$

$$E = \frac{h}{\lambda} = mc$$

where

$$m = \frac{h}{\lambda \, c} = \frac{h}{\lambda \, v} = \sqrt{\frac{\nu^2 \, h^3}{G \, K}} \text{ and } c = \frac{\nu \, h}{m} = \sqrt{\frac{G \, K}{h}}$$

Since $v_1^0 = c$, then light would originate solely in the ee domain. The singularity of the phenomenon, so disconcerting for so long, comes from the fact that we make our measurements directly in the U domain, thereby "jumping" the e domain altogether. If not so, this E would be equal to $m \, c^2$ and the condition $E = \nu h$ would not be fulfilled.

We recognize one mass which is equal to the inertial and gravitational one, since by using the domains it is no longer necessary to use a conventional m_o, but rather that which corresponds to each domain. Consequently, the concept of inertia arises, just as we understand it in the U domain.

The first principle, which states that the total amount of mass in the universe, formed by all the domains, is constant, is related to time. It is not easy to see that it is indispensable that this amount be constant, since otherwise we may be constrained to admit the possibility that time does not have a regular flow; all our laws are based on just such a regular flow of time. Two hundred years ago, Kant wrote the following on this matter:

All phenomena are in time, and only in time as their substratum can simultaneity and sequence be expressed. Thus time, in which all phenomenological change must be considered, is permanent and does not change, and sequence and simultaneity cannot be represented as other than its determinations. Time, however, cannot be perceived of itself. We must look at the objects of the permanence, namely, phenomena, the substratum of which is time in general and is that whereby all sequence and simultaneity may be perceived by our appreciation by the relationship of those phenomena with that substratum. The substratum, however, of everything real, that is, everything belonging to the existence of things, is material, in which everything belonging to existence can only be conceived of as determination. Consequently, that permanent entity, whereby all temporal relationships of phenomena are necessarily determined, is the substance of the phenomena, that is, the aspect thereof which is real, and reach such that, as substratum of all change, it remains always the same. Since this substance cannot change its existence, its quantum in nature can neither increase nor decrease.

A STUDY OF ACCELERATION

The acceleration of a body in U, which, by giving up velocity approaches from its potential field another body, happens with a variation of velocity g. This g may take, among others, the following forms.

$$g = G \, \frac{M}{d^2} \qquad \text{(XIII-17)}$$

$$g = \frac{h \, c^2 \, M}{K \, d^2} \qquad \text{(XIII-18)}$$

$$g = \frac{h \, M}{K \, \mu_o \, \epsilon_o \, d^2} \qquad \text{(XIII-19)}$$

$$g = \frac{h \, c^2}{K} \, \frac{\dfrac{G' \, \mu_o \, \epsilon_o}{\lambda \, v}}{d^2} \qquad \text{(XIII-20)}$$

$$g = \frac{h}{K \, \mu_o \, \epsilon_o} \, \frac{\dfrac{G' \, \mu_o \, \epsilon_o}{\lambda \, v}}{d^2} \qquad \text{(XIII-21)}$$

$$g = \frac{h \, c^2 \, \dfrac{E \, h}{G'}}{K \, d^2} \qquad \text{(XIII-22)}$$

$$g = \frac{h \, \dfrac{E \, h}{G'}}{K \, \mu_o \, \epsilon_o \, d^2} \qquad \text{(XIII-23)}$$

If we take expression (XIII-18), we see that M c^2 is the energy producing the acceleration. There is a position factor, d^2, and h/K is the quotient which shows us the ee:e ratio of the phenomenon, or how many times greater the phenomenon is in e (given by h) with respect to ee. This is somewhat similar to what happens with Coulomb, since we shall see that it is equivalent to expression (XIII-5) multiplied by C/G as a potential quotient.

We now ask why the approach of the bodies is accelerated and why a body does not give up velocity at a constant rate and move at a constant speed instead of doing so in an accelerated fashion.

The answer lies in there being a potential field gradient. A body in a region of the potential field with a value n will have a velocity

$$d\,v_1 = \frac{d\,x_1}{d\,t_2} \, ,$$

and in a displacement, dx, move to a lower potential n-1, whence

$$d\,v_2 = \frac{d\,x_2}{d\,t_2}$$

where $t_2 = t_1$;

$$v_2 = v_1 + \Delta\,v$$
$$x_2 > x_1$$

and Δv = (velocity in potential n) – (velocity in potential n–1), giving rise to the acceleration, although the velocity contribution is constant: giving up the same amount of velocity, it covers greater distances.

On arriving at the potential field value of zero, the acceleration reaches its maximum and the reason for giving up velocity is extinguished.

With bodies in space, this phenomena is reciprocal.

A STUDY OF THE STANDARD KILOGRAM

The problem with the standard kilogram is that it cannot be rendered as a function of other parameters, as, for example, the standard meter, which may be expressed as a function of a number of particular wavelengths.

We shall endeavor to define it as a function of a frequency, a wavelength, and an energy. Bearing in mind that:

133

$$P = m.g$$

$$g = \frac{h.c^2.M}{K.d^2}$$

$$m = \frac{E.h}{G'}$$

1) $E = m\dfrac{KG}{h}$, hence $m = \dfrac{E.h}{KG}$ and

$$g = \frac{h\,c^2\,M}{K\,d^2}$$

$$P_1 = \frac{E.h^2.c^2.M}{K^2.G.d^2} = E.K_1 \qquad \boxed{P_1 = E.K_1}$$

And in Earth:

$$K_1 = 10^7 \text{ cm}$$

$$E = 10^{-4}\,\frac{gr}{sec^2} \text{ hence}$$

$$1 \text{ Kg} = 10^{-4}\,\frac{gr}{sec^2} \times 10^7 \text{ cm} = 10^3 \text{ gr.}\frac{cm}{sec^2}$$

2) $P_2 = m.g = \nu\,\dfrac{KGhc^2M}{Kd^2\,c^4} = \nu\,\dfrac{KGh\,M}{K.c^2.d^2} = \nu\,\dfrac{h.G.M}{c^2.d^2} =$

$$= \nu\,\frac{h^2M}{K.d^2} = \nu.K_2 \qquad \boxed{P_2 = \nu.K_2}$$

$$K^2 = 9.3 \times 10^9 \text{ cm/sec.}$$
$$P^2 = \nu.\,9.3 \times 10^9 \text{ cm/sec.}$$

Relation of time (frequency) to weight:

$$\nu = 1.1 \times 10^{-7} \text{ gr/sec.}$$

$$1 \text{ Kg} = 1{,}1 \times 10^{-7} \text{ gr/sec.} \times 9{,}3 \times 10^9 \text{ cm/sec.} = 10 \times 10^2 = 1.000 \text{ gr}\,\frac{cm}{sec^2}$$

3) $P_3 = \nu K_2 = \dfrac{1}{\lambda}.K_2 \qquad \boxed{P_3 = \dfrac{K_2}{\lambda}}$

$$\lambda = \frac{1}{\nu} = 1.1 \times 10^7 \text{ sec./gr.}$$

$$P_3 = 1 \text{ Kg} = \frac{K_2}{\lambda} = \frac{9.3 \times 10^9 \dfrac{cm}{sec.}}{1.1 \times 10^7 \dfrac{sec.}{gr.}} = 9.3 \times 1.1 \times 10^2 \text{ gr. } \frac{cm}{sec^2}$$

$$= 1.000 \text{ gr. } \frac{cm}{sec.2}$$

Therefore, in accordance with its mass in U, any body in space must have some constants which, multiplied by some parameter, for example, wavelength, frequency, etc, will determine the standard kilogramme for us.

In a comparative manner using, say, the energy between different bodies, will be able to save the standard kilogramme on a third body.

For the moon:

$$\frac{h^2.c^2}{K^2.G} = 10^7 \text{ cm } \frac{d_T^2}{M_T}$$

$$P_L = \frac{h^2.c^2.M_L}{K^2.G.d_L^2} .E = 10^7 \text{ cm.E. } \frac{d_T^2.M_L}{M_T.d_L^2}$$

$$M_L/M_T = 0.44$$

$$d_T^2/d_L^2 = 3.66$$

$$P_L = P_T.(0.44 \times 3.66) = 10^3 \times 1.6 \text{ gr. } \frac{cm}{sec^2}$$

hence

$$g_T/g_L = 6 \qquad g_L/g_T = 1/6 = 0.16$$

$$P_T//P_L = 1/0.16 = 6$$

and for a "i" body:

$$P_i = P_T \cdot \frac{M_i}{M_T} \cdot \frac{d_T^2}{d_i^2} = E \cdot \frac{h^2.c^2}{K^2.G} \cdot \frac{M_i.d_T^2}{M_T.d_i^2}$$

or in general

$$\boxed{\frac{P_T}{P_i} = P_i^{-1}}$$

(XIII-24)

135

Having seen what we have up to now, I don't think it necessarily of great interest to go over matters such as equilibrium, moments, rotation, thrust, work, elasticity, harmonic movement, *et cetera*; it is, however, interesting to underline some significant features, among which would be:

1. Centripetal force

$$\Sigma \; F = m \; \frac{v^2}{R} = \frac{E_u}{R} = \frac{E_u}{distance}$$

This expression relates energy with the gravitational field in U for the circumstance $\overline{V}/\overline{v} = 1$, hence constituting the link between our description of the gravitational phenomenon and that of Newton. This is so insofar as centripetal force does not describe the phenomena per se, but the condition for equilibrium, in the potential field, of two bodies moving through space when there is neither approach nor increasing separation (so-called attraction or repulsion).

2. Kinetic energy

Kinetic energy arises from integrating at a power of unity, later giving an average of $E = m \; v^2$.

$$w = \int_{v_1}^{v_2} m \; v \; dv$$

$$K = \frac{1}{2} \; m \; v^2$$

3. Power

Power is the product of a force by a velocity ($P = F \; v$) and is the same in all domains, since scale change for velocity in moving from one domain to another is unity and because of the conservation principles, which are a consequence of it.

4. The General Enunciation of the Law of Conservation of Momentum

"The total momentum of a system may only be modified by outside forces acting on the system." This is what is behind the phenomenon of the interaction of systems which is observed in the real world—that is that the "outside forces" means another complex system interacting in a certain potential field.

If we say that in the expansion phase of the universe—or a part of the universe—velocities (not accelerations, which, we have already seen, are velocities in a field gradient and originate from some force) tend to distribute themselves, this means that, in a system considered isolated, momentum and its modifications are hidden in that action of transfer.

If we think of this happening in each domain and across domain boundaries we can see the difficulty of grasping the concept.

Chapter XIV

THE POTENTIAL FIELD OF
LIQUID BODIES IN THE U DOMAIN

HYDROSTATICS

Hydrostatics itself, being the study of liquids in repose, is not of special interest to us, but the study of liquids in motion in a potential field, *hydrodynamics*, does concern us.

It is clear that liquid and gas, as states of matter, are not in conflict with the general laws with which we are familiar. Even so, we have some new concepts regarding the structure of matter—or, rather, we are endeavoring to focus on situaions in terms of the structure of matter and our constant K, which is that of the relationship of constants of the potential fields in which phenomena take place in that typical structure of matter. In the light of this approach we shall try to gain, if possible, a better understanding of the phenomena of fluid dynamics.

THE TIDES

If the gravitational phenomenon is a potential field defined in space and brought about when a mass circulates in it gives rise to a reciprocal perturbation of the corresponding region of space when another mass is also moving in it, then in certain circumstances of compatibility or equilibrium (when $\overline{V}/\overline{v} = 1$), the two bodies find themselves in a situation of dynamic equilibrium. One or both of the bodies may be completely solid, liquid, or gaseous, or may be some combination of all three; this last must be the most common case.

In this case, since the bodies interact under those conditions, which are the reciprocal influences with respect to which of the states of matter predominates in the bodies in question?

Let us look at a single body, the earth, with the three states of its mass—solid, liquid, and gas—and the consequencess of the perturbation or distortion produced in its potential field in space having it center at the center of the earth and extending out into space *ad infinitum*. (We shall take the earth, though the situation is the same for any body.)

The most obvious phenomenon resulting from the movement of the moon is that of the tides. We don't detect relative mass movements in the solid phase of the earth's mass, and while the gaseous phase (atmosphere, clouds, vapors), being tenuous masses and in movement anyhow, doesn't show detectable tidal movement, it must be there.

Now, what is the tidal phenomenon? We should bear in mind that there is no attraction of the moon by the earth but an equilibrium situation, which is a function of the energy content of the bodies. This equilibrium situation is, however, not uniform in space but varies each time element, dt, by reason of the movement of the bodies. This gives rise to an ongoing modification of the potential field values at each point in the universe.

Now, when the earth moves and hence causes variations in the potential field values by interference with the variations produced by the moon in its movement, it is logical that

(1) as action, the solid masses reciprocally modify the field and the effect is that they adapt themselves to the field, seeking their equilibrium energy level; and

(2) The liquid ones do the same as the solid ones, but, being liquid, can seek their equilibrium energy level independent of the solid parts of the body. This would explain the tides, but we may note

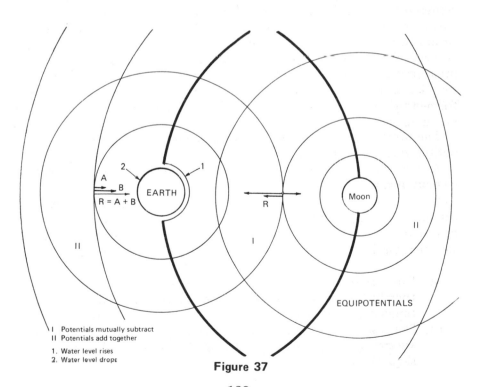

I Potentials mutually subtract
II Potentials add together

1. Water level rises
2. Water level drops

Figure 37

139

that the friction produced takes away energy equivalent to the whole (solid, liquid, and gas) as action.

In other words, when the earth's water responds to the potential gradients through which it passes in space (regularly produced by the moon) it removes field-generating energy from the earth in the same amount (see Figure 37).

STOKES LAW

This law states that when a sphere in a liquid or a liquid flows past a sphere, the resistance R is given by

$$R = 6 \times 3.14 \times T \times r \times V_{rel}$$
where T = viscosity coefficient of the fluid
r = radius of the sphere, and
V_{rel} = relative velocity of sphere in fluid.

This situation, especially if the sphere is falling along the local vertical, represents a potential field that approximates to the ideal gradient. In this phenomenon, it can be clearly seen that in a potential field force is proportional to velocity and velocity alone. This shows that, from a conceptual standpoint, hydrodynamics can be very useful, and since we must prepare for a delicate study of thermodynamics (in the next chapter), let us see what (a) work performed on a system, and (b) work performed by a system means in the passing of velocities between materials and in matter.

The points (a) and (b), above, are complementary, but are they compatible with the third postulate?

As for work performed on a system, obviously the system receives velocity and thus the transfer is fulfilled.

In the case of work performed by the system, from where does it get the energy? If it can be shown that it comes from a lower domain, the postulate is compiled with. If this cannot be shown and the origin of the energy is doubtful, we may suspect either of the following:

(a) The phenomenon is part of the recessive process. If this were so, then both expansion and recession processes can take place simultaneously. Otherwise

(b) The work performed apparently comes from a higher domain. If we trace its origin, we can discover that it does follow the proper sequence.

140

Examples

(a) Work on a system with energy coming from a lower domain:

(1) The burning of fuel; exothermic chemical reactions.
(2) The lifting of a weight, insofar as this, in fact, finds its origin among the phenomena covered in point 1, above.

(b) Work done by the system:

(1) The energy generated in an impact, such as a car crash. We bring about the exit of energy from a lower domain, and to move the vehicles, we also get energy from a lower domain by burning the fuel.
(2) Endothermic chemical reactions. In this case, does the energy from a higher domain go to a lower one? Is this, in fact, a part of the recessive process, or does the energy, in reality, go to a higher domain? For compliance with the third postulate, the last of the above options should be the one that is true. To show this it would be necessary to show that the work of the system providing the energy moves from one domain (*e.g.*, heat from the molecular domain) to another higher one (*e.g.*, atomic or subatomic recombinations at the ee level).

BERNOULLI, VENTURI, AND PITOT

It is interesting to note that Bernoulli's theorem is compiled with when the ratio $\overline{V}/\overline{v}$ is not equal to unity because it concerns a closed system.

$$\frac{P_1}{\rho\, g} + \frac{v_1^2}{2\, g} = \frac{P_2}{\rho\, g} + \frac{v_2^2}{2\, g}$$

In a nonuniform potential field through which there moves a body containing a system of this type, since the g value will be other than that at the surface of the earth, then a different energy will also be necessary to obtain the same pressure. However, we can write the following:

$$p = \frac{h\, c^2\, M\, m\, \overline{V}}{K\, d^2\, \overline{v}} \, .$$

141

Thus, a different $\overline{V}/\overline{v}$ relationship, or one that is continually changing, will give rise to the corresponding variation in the energy contribution. In short, on varying g and the $\overline{V}/\overline{v}$ ratio, equality is maintained, and for this the energy necessary will change:

$$\Delta E = f(\overline{V}/\overline{v}).$$

This is so because the ratio $\overline{V}/\overline{v}$ gives us the factor that must be applied to the force bonding two bodies, with respect to the equilibrium between them.

If the force becomes very large (and the g value very small), the $\overline{v}$ will be large and either add to or subtract from that of the flow, depending on the flow direction, in the proportion given by the equation. This, in the final analysis, will be a variation in the energy consumption of the impeller pump motor (if it should be such a type).

Thus we see again that equilibrium is a function of velocity and energy a function of the moving mass.

Venturi and Pitot are governed by this law; but, now, how does the measuring of velocity affect the actual velocity of the moving body?

The moving body has a certain $\overline{v}$, and the pressure p is employed to determine the velocity of the moving body with respect to the surrounding medium.

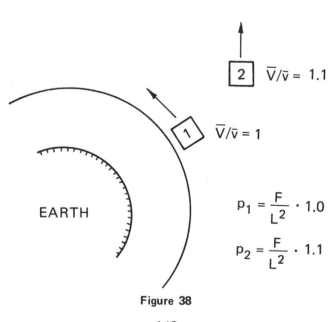

$$\boxed{2} \quad \overline{V}/\overline{v} = 1.1$$

$$\boxed{1} \quad \overline{V}/\overline{v} = 1$$

EARTH

$$p_1 = \frac{F}{L^2} \cdot 1.0$$

$$p_2 = \frac{F}{L^2} \cdot 1.1$$

Figure 38

142

If the $\overline{V}/\overline{v}$ value is large, this would mean that the force necessary is large, and $\overline{v}$ would be small (*e.g.*, $\overline{V}/\overline{v}$ = 30/1 = 30).

If the $V/\overline{v}$ is small, it would mean that the force necessary is small and that the $\overline{v}$ value is large (*e.g.*, $\overline{V}/\overline{v}$ = 30/60 = 0.5). Pressure will also vary in the same way as force.

If $\overline{V}/\overline{v}$ is equal to 1.1 and the velocity is measured in the atmosphere by Pitot, the pressure p will increase by ten percent if we assume a vertical ascent because we measure with respect to $\overline{V}/\overline{v}$ = 1 in a reference orbit of horizontal movement.

The pressure reading indicated by Bernoulli as a velocity value will be exceeded by ten percent.

The fact that, with an increase in altitude, the g value and the pressure of the fluid (the atmosphere) decrease makes the detection of the error produced a little difficult, but it can be done.

Chapter XV

MOLECULAR POTENTIAL FIELD IN THE e DOMAIN

TEMPERATURE AND HEAT

Since a body's temperature is a measure of its relative state of heat or cold, and since heat is a form of energy, amounts of heat, just as mechanical energy, can hardly be grasped at all unless we consider it in terms of the structure of matter and the domains.

Thus it seems that heat is generated by molecular agitation, it belongs to the e domain, and it is transitory insofar as it is a perturbation which subsequently dissipates (or is transmitted) as the molecules give up velocity. Now, what is the basis of this tendency to give up heat if it is not the fact that the heat carried by the mass is greater than what it needs for the equilibrium conditions in the potential field in which it is located? So, if we add energy to increase the mass temperature from 10° to 20° while 10° is the stable situation, the mass will give up the heat corresponding to the 10° temperature difference. This stable situation, of course, is with respect to the surrounding medium, since the stable situation without heat or molecular agitation would be that of absolute zero.

Now, if we are looking at molecular agitation and heat is an e domain phenomenon, what heat may there be in the ee domain? Should it be absolute zero (0° K) or could it be 3° K, the "background energy" of the universe?

If we add the same heat amount to different masses for the same period of time, the final temperatures of the masses are different; the heat or energy is closely related to mass. Now, since the expression, calorific energy, is imprecise, we refer to it more correctly as *internal energy*.

If this energy really is a velocity and this is a force, then heat (or energy) transfer is the transfer of forces—that is, velocities and the relationship between heat and temperature is known as heat capacity (Q/Δ temperature) and specific heat is Q/mΔ temp. By definition,

$$\frac{Q1}{Q2} = \frac{T1}{T2}, \text{ or } Q \sim T.$$

From the expressions $E = mv$ and $Q = mc \, \Delta t$, we should conclude that $c\Delta t = v$, but is this, in fact, the case? In line with the concepts in

144

terms of which we are interpreting phenomena, it is. This is a little strange, just as it was strange to find the relation $m-m_0 = \Delta v$.

Since all of these are manifestations of energy, let us continue in order to see whether, in fact, through the domains, temperature has anything to do with mass and energy, apart from the simple function of indicating the amount of heat contained in a body. We would like to know what temperature really is; and we are in for a surprise.

First, let us recollect that the Dulong and Petit law indicates that, taking a mass unit of the gram-atom, specific heats of metals are almost all equal to 6 cal/at. gr. °C. This leads us to assert that the heat capacity of a metal depends only on the number of molecules, at heat energy (velocity) carriers, and not on the mass of each molecule.

If we agitate molecules with a flame, for instance, there is immediate transfer of excess velocity (perturbation) from molecule to molecule, giving rise to propagation of the heat. Convection, on the other hand, is the propagation of heat from one place to another by the actual movement of the hot substance.

Thus, in the first case, energy is transferred in the e domain, while in the second case, although heat has its origin in the e domain, the mass moves in U.

Radiation is different again, being the continual emission of energy from the surface of bodies. In principle we should assume that radiation comes from all the lower domains.

With an increase in temperature, radiation per unit time increases sharply, being proportional to the fourth power of the absolute temperature (Stefan's Law): $R' = e\, \sigma\, T^4$, where R' is the energy emitted in ergs per second per square centimeter, σ is a constant (5.6699×10^{-5}), T is the temperature in degrees Kelvin, and e is a percentage characteristic of the emitting body and surface in particular.

Thus, if $E = mv$ and $v = \Delta t$, then

$$R' \sim \Delta t \quad . \tag{XV-1}$$

However, where is the mass?

Let us look at the table in Figure 39 and see what we find.

In the ee domain, mass has unit velocity ($v^0 = 1$) and in the eee mass and force become a single confusion.

In the eee domain, temperature, as an indicator of energy, takes on its first value, $T^1 = T$, and "mass does not exist there." In that domain, only energy is measured, and this is why it arrives in U with the fourth power. I don't think this is difficult to believe and it is readily acceptable, none of which stops it being surprising, because

145

DOMAIN	T	E	GRAVITATION*
eee	T	?	R
ee	T^2	m	Kp
e	T^3	mv	K
U	T^4	mv^2	G
UU	T^5	mv^3	G*

*See Chapter III.

Figure 39

(1) it adapts itself to the number of domains the existence of which we have been supposing;

(2) it takes a value of unity where energy really begins to make itself manifest (as velocity in ee); and

(3) it justifies R' not having mass for the calculation in U, as in Stefan's law.

THE BACKGROUND TEMPERATURE OF THE U DOMAIN

In the unified field, we can write

$$R' = \frac{E}{L^2} = \frac{m\,G'}{h\,L^2} = \frac{\dfrac{h}{\lambda\,v}\,G\,K}{h\,L^2}$$

$$= \frac{h\,h}{\lambda\,v\,\mu_o\,\epsilon_o\,h\,L^2} = \frac{h}{\lambda\,v\,\mu_o\,\epsilon_o\,L^2}$$

hence

$$e\,\sigma\,T^4\,(L^2) = \frac{h}{\lambda\,\mu_o\,\epsilon_o\,v}\quad \text{and}$$

$$T^4 = \frac{h}{\lambda\,v\,\mu_o\,\epsilon_o\,e\,\sigma\,(L^2)}\quad \text{and}$$

$$T = \sqrt{\sqrt{\frac{h}{\lambda\,v\,\mu_o\,\epsilon_o\,e\,\sigma\,(L^2)}}}\,. \tag{XV-2}$$

For e = 0.3 and λ = 5 x 10^{-4} cm (for 300 °C), we get

$$\boxed{T = 3 \times 10^8 \text{ gr}^{1/4}/\text{sec}^{1/2}}$$
(XV-3)

With $E \sim R' \sim \Delta t = t_2 - t_1 = T$
and $t_1 = 0$ is $t_2 = T$
and then

$$E = m v = m T = 10^8 \text{ gr}^{1/4} \times 3°K \text{ sec.}^{-1/2}$$

Can we really take 3°K as the universe background temperature, considering background temperature as that lying imminent in the eee domain where basic units, or blocks, by their mutual interaction give rise to forces, and that background temperature is the indicator of the basic force of those mutual interactions?

This may be the case if, at least, the fact that m = 10^8 gr.$^{1/4}$ sec.$^{-1/2}$ measured from U can be explained, with adequate scale changes (which we are not in a position to carry out, but only suppose).

For this, we must remember that in eee mass does not exist and we should take the mass value and dimensions of expression (XV-3), as a coordinatable passage constant, or because it is the result of the number of constants taken to calculate T in expression (XV-2).

We are not in a position to show that m = 10^8 gr$^{1/4}$ sec$^{-1/2}$ since, in the eee domain, mass loses its identity. We can only leave this situation open for study and say that it shows a possible relationship of the unitary T located in eee, with the universe background temperature.

Internal energy is, in essence, part of universal energy. Here we understand universal as covering all the domains. Since energy, however, is what is in the lower domains, and our interest here is to give a proper basis for the third postulate, we should remember that this internal energy arises in U and that we detect how bodies expand and contract and thermocouples develop an electromotive force, *et cetera*.

In Figure 36 we saw that $T^3 \sim mv$, that is, $T^3 \sim C$. Hence

$$T = \sqrt[3]{c}$$

$$T = 3000 \text{ cm/sec} = 3°K$$

Therefore 1° = 1000 cm/sec.

With a temperature increase of 10° the molecular $\overline{v}$ will have a value of 10^4 cm/sec.

If X molecules are involved, each with a value Y which may be vibration, displacement, *et cetera*, Y ($\overline{V}$/molecules) = (10^4 cm/sec)/(X molecules); there is no need to make a scale change between domains since velocities are what is involved.

If this line of calculation is correct, we will have a v value for each molecule to study potential field interferences in e, just as we do in U when, from U, we see the phenomenon as being one of heat.

Internal work is that performed by one part of a system on another part of the same system, and we shall not take it into account.

Heat flow is an energy transfer brought about by a temperature difference—that is, a potential field difference—between two bodies.

Velocity distribution is readily visualized in the passing of heat energy from bodies at high temperature to ones of lower temperature. This itself is part of the tendency of nature toward homogeneity, the uniform distribution of all the available velocity or movement throughout all the available mass.

THE PRINCIPLES OF THERMODYNAMICS

If Q is the energy supplied to a body by heat transfer and W is the energy extracted from it by having to do work, the difference, Q – W, represents the variation in the internal energy of the substance.

Now, the First Principle of Thermodynamics states that $dQ = dU + p\,dV$, since $dV = Adx$ and $p = F/A$. and work transfer in the U domain is performed by variation of volumes, at least in the aspects of heat we are considering.

Another way of expressing this would be "Energy cannot be created nor destroyed," which coincides with our principle of a constant amount of velocity in masses and that between them there is only energy transfer. This approach is broader insofar as it addresses itself to all domains, while thermodynamics, conceptually, only covers e and U.

There are adiabatic transformations, in which the system neither gains nor loses heat, and there are isochoric transformations, in which the system suffers no volume variation and, in U, no work is performed. The energy given to the system is entirely used in increasing the internal energy.

There are isobaric transformations, in which the system stays at a constant pressure. An example would be one in which heat absorbed by each mass unit is the heat of vaporization, $L: Q = m.L$. In this case, again, L is related to a characteristic or mean velocity of the molecules in U_L and V_L, by the relationship $mL = (U_V - U_L) + p\,(V_V - V_L)$.

In molecular theory, the force existing between molecules is conceived of as gravitational but partly, at least, also electrical, all this, however without following the inverse square law with distance. In addition, when the separation between molecules is large there is attraction, and when it is small, repulsion.

The conceptual chaos above has its origin in the historical explanation of the phenomena having been first, with gravitational concepts, then with the laws of gases, and finally with well-based electromagnetic theories.

In the light of the equalizing of the equations of Newton and Coulomb, and taking gravitational and electrical phenomena as unified, it is not difficult to present the different situations occurring in the c domain potential fields which will be governed by the ratio C/G.

This is to say that for an outright gravitational phenomenon, the force constant, F, in terms of the energy, E, will be $1/G$. For an outright electrical phenomenon, this constant would have a value of C/G^2. Undoubtedly, since we consider that between the two there isn't such a clear separation, there must be intermediate values according as the transition from one to the other takes place. If this had, in fact, been previously suspected, we see that the molecular phenomenon provides evidence for it.

All this leads to there being a relationship between the energy (kinetic and potential) of a moving mass and temperature, the indicator of that energy. This could mean that the phenomenon is gravitational, electrical, or transitional, this last referring to a case where we may not apply the equations of phenomena the nature of which has been clearly classified as one or the other.

This is why the use of either formula in particular, when we ought to be using one having an intermediate value, will give erroneous results.

The upshot is that distance is blamed for such results since it wasn't possible to come up with an intermediate potential phenomenon having an intermediate constant which itself is the quotient of two others.

In the case of gases, equations of state are obtained using the laws of mechanics:

(a) A perfect gas is made up of perfectly elastic molecules.
(b) Between impacts, the molecules move uniformly and in a straight line.
(c) The molecules exert no force upon one another, except during impact (sic).
(d) The mean kinetic energy per molecule is proportional to the Kelvin temperature.

Now, since the mean kinetic energy per molecule is given by the expression $\frac{1}{2} mv^2 = 3/2 \, kT$, and remembering that kinetic energy comes from an integration giving a mean as the average of extreme values, if we relate the coefficients, v^2 with T and m with k, we obtain an identity. In this, it is allowable to take the proportionality of v to T, for the

149

reasons stated previously, and also of m with k (Boltzman's constant). This is so because this last is the relationship of an energy per molecule and per mole—that is, it has to do with mass.

Therefore, if we take the equation of state pV = NkT and make a substitution using p = F/A, taking an intermediate value for C/G, calling it $(C/G)^i$, then we get F = NkT/(L). Then, with the value of F in terms of E, we obtain

$$T = \left(\frac{C}{G}\right)^i \frac{h^2 E_1 E_2 \overline{V}}{K^2 N k \overline{v}}.$$ (XV-4)

In this case we have a relationship of T to $\overline{V}/\overline{v}$. This is a relationship between an energy indicator and the velocity contained in the masses of the e domain and the value for each case of the constant of approach/separation (attraction/repulsion).

We can think of this constant taking a value for which there is neither approach nor separation between the two particles; moreover, it must be so. Probably the atoms or molecules of the e domain, in crystals where they are almost stationary in positions fixed in space in relation to their neighbors, are in this situation.

Expression (XV-4) can also be deduced starting from the relation pv = nRT.

THE SECOND PRINCIPLE OF THERMODYNAMICS

The second principle of thermodynamics says that it is impossible to construct an engine such that, functioning periodically, it produces no effect other than that of taking heat from a heat source and converts that heat entirely into work.

Another way of expressing this is that entropy can be created but not destroyed. For this reason we include it as a corollary of the third postulate since, in another way, it can mean that velocities tend to distribute themselves in masses inasmuch as entropy may be taken as a synonym for distribution.

Entropy has been presented as disorder, and, in fact, without disorder, there is no distribution. If there were order, we would have arrived at a situation of final equilibrium; hence disorder is a vision of the whole process of velocity transfer in masses.

For the ee domain, there may be another entropy, since we interpret in the U domain events that happen in e.

If:

150

$$\int_{1}^{2} \frac{dQ}{dt} < S$$

which is why Q/T always increases. How could it have been otherwise? If Q is energy and T its indicator, then when energy circulates between e and U, it produces work in the expansion process of the universe. This confirms that

(1) energy follows the course of lower domains to higher ones, and
(2) entropy indicates the passing of velocities.

Since Q/T ~ S is the quotient between proportional numbers, with a larger Q we have a larger T and larger S (indicating more transfer of velocities). For a constant Q, if Q is large and T smaller, then S is larger (implying much greater transfer of velocities). The comparison is made for different processes.

Because of the above, the most efficient processes are those at low temperature (for example, those of the human body compared to those of a gasoline engine).

Since efficiency encapsulates the value of a loss, let us note one of the most extraordinary consequences, according to the structure of matter: When energy passes from one domain to another, the efficiency is 100%. We find *moto perpetuo*, the perfect machine.

Can this be possible? Let us look more closely.

First, the efficiency in e seen from U is less than 100% (look at any cycle you fancy).

If efficiency increases at low temperatures, it means that it can come as close to 100% as you like, but we can never state flatly that it actually gets there.

Now, let us think about the loss produced in a process in a domain. If the process has an efficiency of 60%, then 40% is the loss "of such a system in such a process in such a domain."

If we then trace the 40%, we will find that it follows the course given in the third postulate. We must be careful, however, that the 40% stays in the domain of the event and 60% passes on to the next higher domain.

In the final analysis, it seems that everything must be due to the elements constituting the masses in each domain being masses that move in a vacuum, and since in a vacuum there are no losses as we normally understand them, everything must work as in *moto perpetuo*.

If we had a workshop and were requested to make such a perfect machine, being given the key on how to do so and the means to go ahead, then we would consider the following: a mass I made up of a vacuum

151

and constituted of rotating systems supplied with a starting force. This force would give a resultant of I that, with another similar mass II, would form a system that we would call from a higher domain and according to a condition for equilibrium, they would form a system rotating in the vacuum, *et cetera*, forming a new mass.

Where could there be any loss?

Thus arises the question of why the energy of the lower domains ascends: Could it not be that that energy feeds the losses that we don't find but that do exist?

If this were the case, the universe would expand, never to stop, and turn around and everything would follow a straight and defined path with no return.

On thinking, however, of who put this feed energy in and with some clues gathered from astronomy, we arrive at the conclusion that the system does turn back, it is alive and palpitates, and that it has an expansion and a contraction phase. We don't know where it's going, or whether there are time or spatial limits, but we do conclude that the machine without loss, the *moto perpetuo*, is not formed only with the constructive stage as we indicated above, but that it follows a dynamic self-feeding process which can only terminate in a state of absolute equilibrium of masses and velocities in all domains in the final and definitive stillness.

This, of course, is a state that, like infinity, we cannot imagine, because it would be a senseless end; hence the opposite is, for us, more acceptable. This would be a pulsating universe which always renews itself.

I think that to understand such a mechanism is the ultimate marvel of nature.

A METHOD TO QUANTIFY HIDDEN VARIABLES IN THE UNIFIED FIELD

It is logical to ask at this stage whether some of the unified field expressions may be used, not only to interpret nature but also to solve problems and advance on a firmer basis in the knowledge of the constitution of matter.

We would specially mention the hidden variables as variables of ignorance. Thus if we take molecular movements as an example, we can enunciate what, up to now, we have possibly glimpsed as a practical application of the new relationships—namely, three lines of calculation.

Interpolation of Constants

We can take i out of the expression (XV-4), obtaining thereby $(C/G)^i$ = X, where X is the constant for the phenomenon under the conditions given.

We can say in general that for calculations in a potential field of any domain, with the constant unknown, it may be determined from two known or extreme constants by means of adequate mathematical interpolation.

The Quantification of a Phenomenon Not Corresponding to a Given Expression

If in an expression we have a variable that is a function of another and we need to determine, in the phenomenon concerned, a parameter that does not appear in the expression given, we can make the change to calculate it. If we have, for example, $E = mc^2$ and we want to know the wavelength in a simple isolated phenomenon as a function of G, we may write

$$E = \frac{G' \mu_o \epsilon_o}{\lambda v h}, \text{ hence } \lambda = \frac{G K \mu_o \epsilon_o}{E v h}.$$

To Obtain a Constant in Terms of Others

An example of this would be where Coulomb may be written as

$$F = N \frac{E_1 E_2}{d^2} \left(\pm \frac{\overline{V}}{v} \right), \qquad \text{(XV-5)}$$

where

$$N = \frac{C h^2}{G^2 K^2}.$$

Note: In the tables at the end of the book, there is a list of elementary particles in which the mass of both the photon and the neutron masses are given simply as a question mark on the reasoning that we may no longer ascribe a zero value to a mass.

Mass is expressed as energy (Mev), which, with the advance of knowledge, must also be separated into mass and velocity components.

Chapter XVI

ATOMIC POTENTIAL FIELD
IN THE e DOMAIN

ELECTRICITY AND MAGNETISM

Having covered some heat phenomena in the e domain in the previous chapter, and left out others such as acoustics for lack of immediate interest, we now turn to electric and magnetic aspects.

In the e domain we will look at high-energy phenomena where $i = 1$ in expression (XV-4). Also, since we now have a method to quantify (hidden) variables, we will pass quickly through the various aspects so as to put the basic notions on a firm basis.

In the Appendix we will see some cases treated as examples worked out in more detail.

We shall begin by noting that the kilogram unit in the U domain is equivalent to the Coulomb unit in the e domain. This, of course, requires the application of the appropriate scale change and this would appear to be that for distance (equal to that of time), thus:

$$r = \frac{C}{G} = 1.35 \times 10^{20} \frac{Coul^2}{Kg^2}$$

$$\frac{1}{r} = \frac{G}{C} = 0.74 \times 10^{-20} \frac{Kg^2}{Coul^2} \ .$$

The product $(r \times 1/r)$ equals one and is equivalent to the scale change for velocities; then

$$\frac{Kg^2}{Coul^2} = 1$$

$$\sqrt{Kg^2} = \sqrt{Coul^2} \ .$$

The kg = Coulomb equivalence is fulfilled if C and G are measured from U or if we wish to measure, from U, the C value in the e domain. In this case, we must apply the indicated scale change.

With this condition, the Coulomb is the unit of force, in the e domain potential field, which is equivalent to the kg function in the U domain.

From expression (XV-5) we can say that the force of approach or separation of masses bonded in an electric potential field is directly proportional to the product of their energies and inversely proportional to the square of the distance.

The same could be done for gravitation, with $N^* = h^2/Gk^2$ and, in general,

$$\frac{\partial F}{\partial t} = N^* \frac{\partial (\overline{V}/\overline{v})}{\partial d} \quad . \qquad \text{(XVI-1)}$$

If we have a stellar body or an atomic particle, we may represent the field in spherical coordinates, or simplifying, on a plane surface passing through the body's center, in polar coordinates.

The superimposing of field squares will give rise to instantaneous point vectors, in the field, composed of the interfering fields. By convention, if their directions indicate the concave part of lines or equipotential surfaces, then they are positive and cumulative. For opposite directions, they are subtracted (in showing the convex part, they are negative). In the system, we call the final vector the field resultant for a certain t and position.

We now find ourselves with the question of whether, in the unification of the fields in physics—since we now have but one—we can define the instantaneous position of particles. In other words, must quantum mechanics remain statistical, or is there another solution?

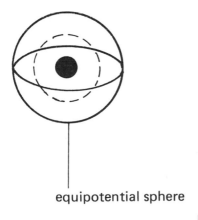

equipotential sphere

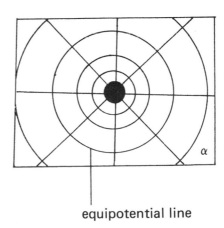

equipotential line

Figure 40

The answer is definitely no; statistics must stay because the time scale change renders it impossible to measure, in U, what is going on in e. Practical necessity thus demands the ongoing use of statistics. It is, however, very simple to work with e domain time (distances) and follow a mechanics like that of the U domain.

Let us look more closely at this. Compared with the situations of U, the phenomena of the e domain are high velocity, since velocity is a single entity. If we want to determine instantaneous positions at a given time, we employ scale changes and obtain values since we know the laws governing the relationship between the particles in their potential fields, whether they be isolated or interfering. The isolated values will be very small, so much so that they will be almost useless from a practical standpoint. This is especially the case if we bear in mind that, in e, we are working with a large population of simple data, and statistics gives good results under those conditions. Even so, using expression (XVI-1) and a large number of particles, both results should coincide and, in this, we are able to make quantum mechanics more deterministic.

As for the relationship q/m (charge/mass) in the e domain, we must remember that q = charge = $m\bar{v}$, and mass is said to "increase."

$$\frac{q}{m} = \frac{q}{m_o} \sqrt{1 - \frac{v^2}{c^2}}$$

In fact, we are measuring something as charge that may no longer be referred to as such where $m_o = m_e$; hence

$$\frac{q}{m} = \frac{m_e \bar{v}}{m_e} = \frac{m_e \bar{v}}{m_e} \sqrt{1 - \frac{v^2}{c^2}} = \bar{v} \times N^r.$$

Thus, instead of a mass increase, we have a velocity variation since, with q, we get a value for $m_e\bar{v}$, a product; and if m is constant, what varies is $\bar{v}$.

INTERFERENCE OF POTENTIAL FIELDS

We have been determining the assimilation of the different laws of phenomena into the basic energy expressions and we shall continue to do so, not only to verify that the assumptions contained in the postulates are correct but also finalize this stage and pass to the next, which must involve the following: The determination of the equilibrium and non-equilibrium conditions for the phenomena of all the domains on the basis of the unified field equations.

For this, we must also review the fundamentals for each task, and first, we shall look at potential fields.

In potential fields, the potential at a point is defined by the quotient of the force required to remove a reference unit from the equilibrium. For example, in an electric field it is the quotient of the force per unit of charge q.

The field direction is the direction of the force. A line of force is an imaginary line drawn such that its direction at each point is the same as the field at that point.

The number of lines of force may be used to obtain the direction and magnitude of the field. Gauss's theorem says that the net number of lines of force passing through any closed surface toward the outside is equal to the net positive potential enclosed by the field.

As a result, the field outside of a uniformly charged conducting sphere, for example, is the same as though all the charge were concentrated at the center.

If the electrical phenomenon is similar to the gravitational one in all the domains, but of high energy, and the gravitational one is considered as electrical but of low (or minimum) energy, then to study conditions of equilibrium or nonequilibrium, we should probably not use the practical units (volts, *et cetera*) since the real meaning of what we are doing would be lost from sight.

The same would apply to magnetic phenomena where we should (as we have been doing) adopt, for example, only the CGS system, although subsequently the practical system would be applied for obvious practical reasons.

All phenomena, then, should lend themselves to being represented in a system of squares in the field in which they take place or in the interference produced by the interaction between fields. A simple example is that of Figure 37 and another would be the displacement of a neutron. or an electron in the e domain.

With the proper scale change, this would be absolutely deterministic, although subsequently statistical methods would be used for a large population or for an uncertain location in time and space if the domain-passage scale change is dispensed with.

In regard to the energy-conservation principle as a corollary of the third postulate, it may be said in general that the work carried out is equal to the sum of the variation of kinetic and potential energy.

The potential gradient in a field is equal to the quotient of the variation of potential with distance with respect to direction, whence it may be deduced that volt/meter = Newton/Coulomb. Let us compare similar equations of electrical conductivity:

157

$$i = \sigma\, A\, \frac{Va - Vb}{L} = \sigma\, A\, \frac{\Delta V}{L}$$

with that of heat in the stationary state:

$$H = KA\, \frac{t_2 - t_1}{L} = KA\, \frac{\Delta T}{L}\,,$$

where i = electric current, H = heat current, K = heat conductivity, σ = electric conductivity, $\Delta V/L$ = electric potential gradient, and $\Delta T/L$ = temperature gradient. In domain e, then, we would have:

$$\Delta V = \Delta T = \Delta \overline{v} \qquad\qquad \text{(XVI-2)}$$

This is the starting point for developing a detailed analysis of the phenomenon in the potential field in question.

The connection between the phenomena may be given by Joule's law, since the amount of heat produced per second is directly proportional to the square of current intensity.

The goal here is a detailed description, in the potential field of each domain, of all the phenomena that take place there; and it should be found that independent of their manifest nature in U (heat, electric current, radiation, *et cetera*) they have a behavior pattern, just as we find with phenomena in U, having the same universal laws. We should also find that the results of such phenomena manifest themselves in the next domain in a manner, which can be translated by passage constants, that each domain determines the characteristics of the following one and is formed by that below it as in the Russian Doll Syndrome in regard to the structure of matter and the three postulates and five fundamental corollaries.

In the case of a magnetic field (and an electrostatic field), it is said to exist if a force is exerted on a movable charge passing through it. Thus we find that magnetism refers to a distortion of a field originating in a mass moving through it.

To analyze this in e, in the unified field, is not easy, but it is perfectly determined. To do so, the value and direction of the magnetic field at a point must first be determined, taking as data that corresponding to the movable charge that creates the field, and then the value and direction of the force exerted on a movable charge in a given field must be calculated.

The force acting on a movable charge in a magnetic field is proportional to the value of the charge:

158

$$F \sim q \sim \overline{v}$$

with $B = \dfrac{F}{\overline{v}'} = \dfrac{F}{q \, v \, \sin \theta}$ will be (XVI-3)

$$\boxed{\overline{v}' = \dfrac{F}{B}} \quad \text{Wb/m}^2$$

If the weight of the electron is of the order of 9×10^{-30} Newtons and the magnetic force on an electron from a field of $B = 10$ Wb/m^2 is of the order of 4.8×10^{-11} Newtons, we see that the gravitational force (even if we compare different domains) is negligible compared to the magnetic one.

An interesting case showing that energy is carried by mass, by means of velocity, is that of the mass spectrograph. Here ions of different masses trace out different trajectories, the radii of which are directly proportional to the mass of the ions in question, although all the ions have the same velocity. The $m\overline{v}$ value, or kinetic energy, is measured, and different masses are carrying the same $\overline{v}$.

If mass were constant and velocity were to vary, the products would be the same, and velocities—that is, energies—could be compared instead of masses.

Chapter XVII

SUBATOMIC POTENTIAL FIELD IN THE ee DOMAIN

LIGHT

We shall study the ee domain taking what is, as far as we are aware, a characteristic phenomenon, namely, light. If we assume that light and other radiations originate in ee, there will be a certain compatibility and explanation of their behavior and characteristics.

On this basis, we recollect that, in support of it, light, for example, is emitted by a molecule of gas on returning to its normal state after one of its outer electrons has been separated or elevated to a higher energy level.

For ease of explanation, in Figure 41 we suppose that the radiation is to be dispatched from the nucleus, although we could consider it as coming from the electron itself. In (a) the force F is incident and adds energy to the fields of the particles A and B, thereby moving B to the position B'. In (b) once the action of F has disappeared, there is a velocity transfer and a movement from B' back to the equilibrium position.

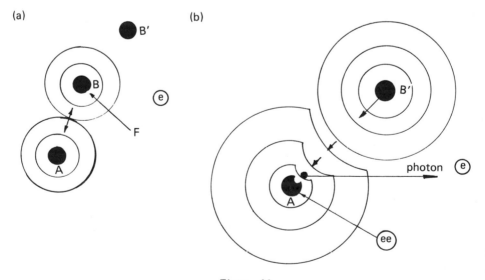

Figure 41

160

This distorts the equipotential line near the surface, A, and this force at this point pulls out what we could refer to as a photon, the same as happened with the tides.

If we accept domain ee as the origin of light, we are in a position to build the theory of light in the unified field on this basis.

Light has a tiny mass and possibly a rotating element which generates an independent field seen in the U domain as a wavelength. This mass is the first mass unit or model and is repeated successively in the following domains.

The velocity is therefore a basic velocity, and we may suppose it to be the minimum velocity allowing escape from the ee domain, passing through the e level to arrive at our level, the U domain.

Its energy is $E = m\overline{v}$, with $\overline{v} = 1$; or $E = \nu h = m1 = m$. Hence $m\overline{v} = \nu h = m$; this means $m = \nu h$. Hence

$$\boxed{\nu = \frac{m}{h}} \, . \tag{XVII-1}$$

It would thus be more correct to write E in general as $E = (m/h)$ GK. This would mean that the frequency, related to the rotation of the quasi-tron (this being what we have chosen to call the supposed rotating element or satellite), is equal to the mass divided by h. This, in turn, means that "one" rotation is a fraction of the central mass; hence, for reasons of mass, force, and energy equivalence, and in regard to the spatial equilibrium in ee, h is a value relating frequency and mass (or energy) which determines whether a particular phenomenon can leave the ee domain. In other words, the F shown in Figure 41 will produce a photon provided the field distortion produced causes, in the interior of A (ee domain with its planetary systems), a planetary system from that interior to detach itself because the tension acting upon it is greater than the tension of its stable situation. This tension of stability is an energy with the value m. If this m, on being divided by h, gives a value less than that of F, then detachment will take place.

I think this covers the matter; however, to orient ourselves better, we can look once again at the phenomenon's passage between domains. We can use the illustration in Figure 42 where we can see that if the photon is from the ee domain, it is made up of units from the eee level! This is where the difference between mass and energy becomes blurred; energy arises from a mutual action between basic units that form the proton. The proton, therefore, is the first unit of mass (as we know it in U) made up of such blocks.

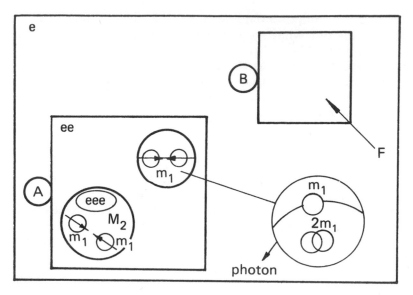

Figure 42

The F looks as though it is never so small or instantaneous to cause the exit of a single photon only, and if it were, we would have some difficulty in apprising ourselves at all because of the minuscule size of the phenomenon. The photons come out as a train, one behind the other. The wavetrain, however, contains photons of different energies and these can be classified, in the case of light, by passing them through a prism.

We could have thought that this mix would finalize in some uniformity where all the photons would have the same total velocity ($\overline{v}$), which would be the sum of C plus the velocity of the satellite.

It would appear, however, that in a wavetrain, the units, because of their ongoing displacement, don't have a chance to make this exchange of velocities.

The problem of the different frequencies is still with us, though, and in Figure 42 we see that the photon must therefore comply with one of the following cases:

(1) nuclei of equal mass and satellites of lesser and different mass, having more velocity in the equilibrium state;
(2) the inverse of (1), above; or
(3) binary systems with components of equal mass.

The first option seems the most plausible insofar as it allows for the formation of variants which we would detect in U as radiation of different types and not necessarily luminous.

If we choose the last option (3) for any reason—mainly elegance—then we cannot classify light as being within the overall category of radiations.

Thus we have the photon nucleus and satellite as the basic building blocks of which matter is made. As for indivisible units, if, for example, the quasitron were a simple unit, it would be that unit that, in mutual action with other similar ones, in eee (or where, in the last analysis, it was once stored) gives rise to forces.

If it is a binary system having a different mass for each frequency, the whole would move at a velocity of c, but each element in orbit at a higher velocity.

Although velocity for us is unitary, down in the lower domains, it is multiplied by such mass, thus obtaining energy. The excess of energy transmitted by the battery (the eee domain) and that begins to form a first more complex mass in ee (which mass, in U, we call radiation), is carried by the nucleus/satellite set, forming a potential field like that of the electron/proton or earth/sun case.

Such excess is quantified by or proportional to the satellite's revolutions.

THE SEPARATION OF COLORS

Let us now look at the separation of colors in a prism, as in Figure 43.

Each photon with its specific energy enters the potential field formed by the atoms in the crystal lattice. These atoms form a potential field almost in equilibrium and the photons classify themselves by energy, insofar as the photons will encounter different field values depending on where they enter the field.

In Figure 43 E_i means energy, proportional to the frequency of each photon; the F_i values are the field resultants toward the atoms A_i: the R_i values are the directions of the energies classified.

The phenomena of refraction and reflexion, as also those inherent to all the other radiations, are explained in the same way.

Double refraction, collimation, and suchlike are phenomena that must be carefully analyzed so as to get to their pure form, eliminating simultanaeity, recomposition, and similar phenomena which do not allow the basic phenomenon to be seen, owing to their complexity in some circumstances.

In Figure 43 there are several spectra for a simple passage through a crystal, while in Figure 44 we see how the passage of the photons through the whole mass of the crystal finally separates the energies, giving a single spectrum.

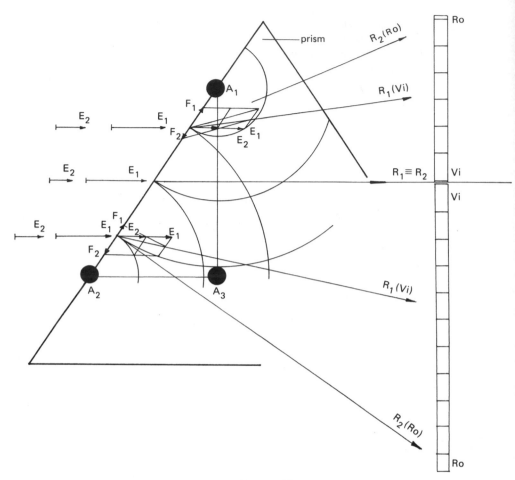

Figure 43

Note that in the case of Figure 41 we have an explanation of the detachment of an isolated body as a consequence of potential fields. We don't say *action*, but *consequence*.

In the void, if there were to exist a single body, its potential field would extend from its center to its surface.

If, by some magic, there appears a second body, an equipotential network has to be set up and lines of force drawn to determine the relative energy conditions between the two bodies. Then it is that we have a potential field extending beyond the surface of the body standing alone.

If we have many bodies an intricate interference between the potential field networks of all the bodies in each domain is formed.

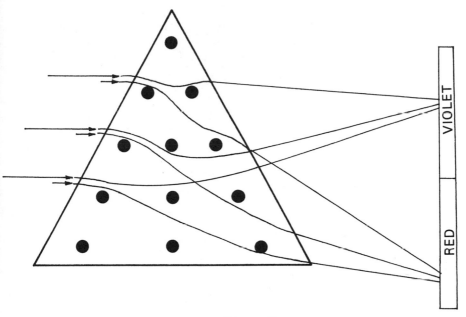

Figure 44

A MASS SHIFTS, OR "SOMETHING MOVES"

In figures 43 and 44 we have an explanation of the deviation of light rays, and it was unnecessary to mention attracting electric charges between crystal nodes, electromagnetism of light, or anything similar. In the unified field there are only potential fields as a result of a single phenomenon—"A mass is moving"—and the uniformity of laws (equations) that arise from this "action."

In line with the criteria mentioned in regard to matter and movement, we can summarize even further, in terms of the maximum condensation of such criteria: "Something is moving."

Let us look at the other forms for the expression of energy, E, by means of frequency and Planck's constant, the same as we did for g and F, thus to arrive at some conclusion as to what that diversity of expressions may mean.

λ	$\dfrac{h}{m.v}$	$\dfrac{G'}{c^2.m.v}, \sqrt{\dfrac{K.G}{m.v^2.\nu}}, \dfrac{KG}{c^2.m.v}$, etc.

165

ν	$\dfrac{E}{h}$	$\dfrac{G'.m}{h^2}$, $\dfrac{c^4.m}{KG}$, $\dfrac{K.G.m}{h^2}$, etc.
h	$\lambda.m.v$ $\dfrac{E}{\nu}$	$\dfrac{K.G}{c^2}$, $\dfrac{g.K.d^2}{c^2.M}$, $\sqrt{\dfrac{G'.m}{\nu}}$, $\dfrac{G'}{c^2}$, $\sqrt{\dfrac{K.G.m}{\nu}}$, $\dfrac{G'.\mu_0.\epsilon_0}{E.\lambda.v}$, $\mu_0\,\epsilon_0\,KG$, $\mu_0.\epsilon_0.G'$, etc.
E	$m.c^2$	$\dfrac{m.G'}{h}$, $m\dfrac{K.G}{h}$, $\dfrac{G'.\mu_0.\epsilon_0}{\lambda.v.h}$, etc.

The conclusion is that they are innocuous. They serve us in making calculations, and one day they will explain phenomena completely.

Let us see this with an example. Let us suppose we wish to measure the air temperature at point A (Figure 45) over a roadway with traffic at 20 km per hour filling it completely. There is a wind of 2 km per hour in the direction opposite to that of the traffic, the air has a temperature of 32° C, and the mean engine temperature of the vehicles is 70° C. There could be other variables, such as humidity, direct action of the sun, temperature of the road surface, *et cetera*.

The value of the temperature at the point A is the result of a very complex phenomenon.

When we finally have the temperature, we know it is a function of movement of hot air, heat generated by the engines, reflection of light from the ground and its absorption by the atmosphere, *et cetera*. Each particular phenomenon has its law, and for each there is a constant.

Let us imagine, so as not to extend the process, that we already have the total expression for the value of the temperature at A, as a function of many parameters and five constants. We may put those five as a single final constant, remembering the calculation of g. Since we have absorption, reflection, radiation, and similar constants in our expression, if we are asked whether we can describe the entire phenomenon, with separation of constants, given the value measured at A, we would reply that it is possible to do so.

If we are asked for the constants and the final one as a relationship of them, we would also be able to do it. If, however, we are asked to determine the velocity of the wind if the engines were to generate more heat, we would be in something of a spot, because to reply we would have to involve the entire atmosphere of the planet in a localized phenomenon.

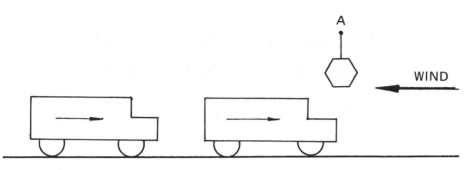

Figure 45

The same thing happens with the series of different unified field expressions for a particular phenomenon, including the constants. In this case we have the added difficulty that we don't have a very good idea of the reciprocal actions shaping a phenomenon. We therefore have the general expression with several constants, the resultant of which is a single one.

This explains why, in U, there is a gravitational constant, or electromagnetic constants in a frequency expression. They simply take part in or have something to do with, the total phenomenon.

The constant of gravitation, for example, although we see no immediate relationship, has to do with the temperature value at A, since the heavier the vehicle, the greater the engine revolutions required to move it, and, in turn, the greater the heat radiated to the atmosphere.

Chapter XVIII

ELECTROMAGNETIC POTENTIAL FIELD IN THE e AND ee DOMAINS

We consider those phenomena that are now classified in the U domain as being of atomic or nuclear physics as being of the potential(s) of e or ee or both domains combined.

In order to get some idea about such phenomena, let us remember that a neutron, within ten minutes of leaving its nucleus (an equilibrium system where the nucleus is not saturated because of a sufficiency of electrons rotating about it), will disintegrate into a proton, an electron, and an inconceivably small neutrino. Should the neutron break up, it produces an impressive amount of subnuclear particles called *mesons*.

BALMER'S CONSTANT

In these domains, phenomena are not well known and less again if they happen in one domain, and we receive them as a manifestation from the other. For example, let us suppose, although we detect something from e, that it in fact originates in ee.

One way or another there are certain uncertainties and the most we can do is to try to situate them in the unified field context. If this situating were to be somehow conflictive, it would constitute a serious embarrassment and cease to be a unification.

The idea here is to try to get Balmer's expression in the form of a potential field equation following the form of Newton and Coulomb. Let us see what conclusions we may arrive at.

We shall follow a deductive process. In Figure 46 we see a sketch to show the problem.

m_q = mass of the quasitron
m_f = mass of the photon
$\overline{V}$ = total photon velocity
$\overline{v}$ = total quasitron velocity
λ = amplitude of the perturbation
ν = frequency or revolutions per unit time (in which mf covers a certain distance)

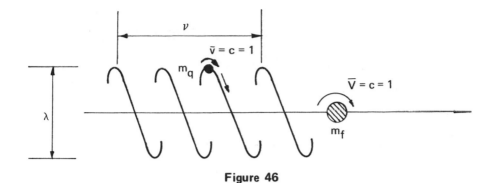

Figure 46

In that domain, $c = 1$ unitary velocity, and for equilibrium, with $m_q \overline{v} = m_f \overline{V}$, the ratio $\overline{V}/\overline{v}$ must be unity. Hence, we have

$$F = \pm\ K' \ \frac{Mm1}{d^2} \ , \qquad\qquad \text{(XVIII-1)}$$

where K' would be a supposed ee domain gravitation constant. We now have Balmer's expression:

$$\nu = Rc\ \left(\frac{1}{2^2} - \frac{1}{n^2}\right) \quad .$$

We realize that it is a quite incomplete expression of the phenomenon. We have to split it in order to compare it with Expression (XVI-1). In the first place, it doesn't express the value of F. For this we need the mass, and we can then write

$$F = \nu\ m_q\ m_f\ 1 = Rc\ \left(\frac{1}{2^2} - \frac{1}{n^2}\right)\ m_q\ m_f\ 1.$$

The bracket indicates a difference between two equipotentials (see Figure 47), the greater of which has a value of one quarter.

We don't know where this difference came from, nor what it may mean. It has, however, served to arrange Balmer's expression and other similar ones and is perfectly acceptable, since we take the bracket as a $1/d^2$ term, thus giving it meaning.

$$\frac{1}{d^2} = \left(\frac{1}{4} - \frac{1}{n^2}\right) \quad \text{and}\ F = \nu\ m_q\ m_f = Rc\ \frac{m_q\ m_f}{d^2}$$

169

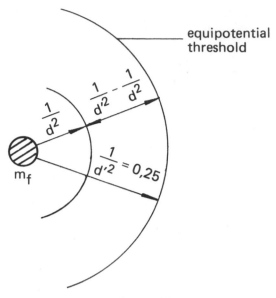

equipotential
threshold

$$\frac{1}{d'^2} = 0{,}25$$

$\frac{1}{d^2}$

$\frac{1}{d'^2} - \frac{1}{d^2}$

m_f

Figure 47

Hence

$$v = \frac{Rc}{d^2} \left(\sec^{-1} x \frac{1}{d^2} \right).$$

Conclusion: Must be $\overline{v} > c > 1$ and $\overline{V}/\overline{v} = \dfrac{c}{c + \Delta c} < 1.$

Then there would be no equilibrium, and it is as follows: $\overline{V} = c$ and $\overline{v} = c$ and $c = 1$, whereby we arrive at the following important conclusion: C is a maximum velocity not exceeded by the quasitron, since, in the ee domain, C is the unitary velocity. If this is the case, why does the nucleus not lose its rotating element if the nucleus itself also has the maximum allowed velocity?

The explanation is that the condition for both to move simultaneously at a velocity of C would be complied with by a binary system where $m_f = m_q$ and $\overline{v}_f = \overline{v}_q = c$. This would mean that what is known as a photon would take on the aspect of the illustration in Figure 48.

This double helix of a binary system reminds us immediately of the DNA structure. If all this were the case, it would seem as though, already in the ee domain, this layout is becoming apparent.

170

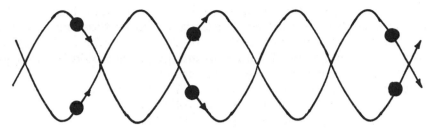

Figure 48

We would then have an initial connection between the inorganic and organic worlds in the unified field. This is so in spite of the series of objections that may be raised, the critical study of which is the way to clearer overall concepts.

The Rc value must represent the gravity constant in eee, seen from ee, and since $c = 1$, this would be equal to R.

SUMMARY OF THE CONSTANTS FOR ALL DOMAINS

$G' = 5.9 \times 10^{-5}$ gr cm^4/sec^3

$G = 6.7 \times 10^{-8}$ cm^3/sec^2 gr

$K = 8.84 \times 10^2$ gr^2 cm/sec

We define $K_p = K/R$ because $K = K_{own\ (p)} \times R$; then:

$K_p = 8 \times 10^{-7}$ gr^2 cm^2/sec

$R = 1.09 \times 10^9$ cm^{-1}

Kp is an unknown constant that we were in need of, being the ee resultant seen from e and that, now that we have R, we can write down. If we now take arbitrary C_i passage values, then:

$$(a) \quad (b) \quad \left.\begin{array}{l} R \\ K \\ K_p \\ G \\ G' \end{array}\right\} \quad \begin{array}{l} C_1 \\ C_2 \\ C_3 \\ C_4 \end{array} \quad \begin{array}{l} C_1\ R = K \\ C_2\ K = K_p \\ C_3\ K_p = G \\ C_4\ G = G'. \end{array} \qquad (XVIII\text{-}2)$$

The internal passages, types (a) and (b), must exist, but they don't get as far as the U level. In U we will always have G, being able to separate out in the unified field, according to the domains, with R, Kp, K, and G. This would mean that all the domain passage constants would be the R of Balmer, the K (and K_p) of the subscript, and the G of Newton, all intrinsic to the respective potential field. The Coulomb, Planck, Boltzman, permeability, and similar constants would be those marking particular energy levels but not intrinsic to the field resultant of a domain in the equilibrium state of minimum energy. Thus would the location of interdomain constants appear to be closed.

Starting out from the expressions (XVIII-2), the passage values between the constants C_i would have the following values: $c_4 = G'/G = K$; $c_3 = G/K_p$; $c_1 = K/R = K_p$; and $K_p R = K$. Then $c_2 = K_p/K = K_p/K_p R = 1/R$; and $R = 1/c_2$.

And to make the total passage:

$$c_1 \, c_2 \, c_3 \, c_4 = \frac{K}{R} \, \frac{K_p}{K} \, \frac{G}{K_p} \, \frac{G'}{G} = \frac{G'}{R}$$

$$\boxed{G' = R.c_1.c_2.c_3.c_4} \qquad \text{(XVIII-3)}$$

Also:

$$C_1 \, R = K$$

$$\frac{K_p}{c_2} = K, \, c_1 \, R = \frac{K_p}{c_2}, \, c_2 = \frac{K_p}{c_1 \, R} = \frac{K_p}{K}$$

$$K = \frac{K_p}{c_2},$$

and since $R = 1/c_2$ we will have

$$\boxed{K = K_p.R.} \, .$$

With this, we can make the following table.

DOMAIN	CONSTANT	VALUE TO PASS TO LOWER DOMAINS
U	G	$G' = G \, K$
e	K	$K = K_p \, R$
ee	K_p	unit/eigenvalue
eee	R	?

172

Let us check whether it is true that the quotient of the extreme values (G'/R) is equal to the product of the passage values of all the lower domains:

$$\frac{G'}{R} = c_1 \ c_2 \ c_3 \ c_4$$

$$G' = Rc_1 \ c_2 \ c_3 \ c_4 = R \ K_p \ \frac{1}{R} \ \frac{G}{K_p} \ K$$

$$= GK = 5.9 \times 10^{-5} \ gr \ cm^4/sec.^3$$

We shall make an interpretation of what we are deducing from the domains, and in this connection we will consider the table in Figure 39 rearranging it in the following way:

Domain	E	T	Gravity	Dimensions	Interpretations
			G'	$gr.cm^4/sec.^3$	Mass x 3 veloc. x length
U	$m.v^2$	T^4	G	$cm^3/sec.^2gr$	1/Mass x 2 veloc. x length
e	$m.v$	T^3	K	$gr.^2cm/sec.$	2 Masses x 1 veloc.
ee	$m.1$	T^2	K_p	$gr.^2 \ cm^2/sec.$	2Masses x 1 veloc. x length
eee	?	T	R	1/cm	1/length

Figure 49

E and T are energy-state indicators and hence must be related to the gravitation resultants in each domain. Let us see what we get. First of all, everything we can deduce should serve in the beginning to dig out what goes on in the eee domain and what is there at all.

In the eee domain, we have the inverse of a distance, as though the dimension of a displacement were somehow suspended. If there is mutual action of the two basic blocks, the dimension applicable is or should be that corresponding to a situation of suspense such as to produce a displacement in the next-higher domain. Also, on jumping to the next domain, ee, we find two masses (see the "Interpretation" headings of Figure 49, a velocity, and a distance.

173

If this (the ee domain) is where light is born and it has its origin in two domains, the two masses may be applicable. In this domain, velocity is the unitary velocity and distance must be the manifestation of the phenomenon coming from eee, where it was distance "suspended," as it were. This is a phenomenon that is a bit awkward to picture clearly and wherein lies the secret of the basic structure of mass and energy.

If this were the case, K_p would be the gravitational constant of the potential field of photons of light.

In the e domain there are two masses, which we take as being that of ee and that of e where they are properly formed, and a multiple v of the unitary velocity taken in the lower domain.

In U, all the foregoing has to be adequately added and we put the whole together and call it G'.

If we were to take G we would be considering the U domain gravitational field constant but not the mass itself or analyzing the structure of mass via interpretation of dimensions.

Thus we have a mass, which is logical for the reasons stated, three velocities—coming from e, ee, and that of U itself—and a distance.

This distance is disconcerting. I think it must have something to do with what goes on in UU. In other words, all the dimensions of U, including this distance, will determine the dimensions of UU.

The constant R reduces the dimensions of K_p to a distance giving K the dimensions of a velocity and a mass taken twice.

The K is also needed so that G' will have a normal mass in U ($G \sim$ gr/l).

With what we have seen up to now, we can try out a possible structure of mass in eee, as in Figure 50.

We can reasonably suppose that in eee there are basic blocks of matter and that they are equidistant from each other, although in eee this does not carry any real meaning. Neither does it have meaning to speak of "energy," but even so its presence generates a potential field whence there arises the constant R, seen from e and which we, in U (having passed constants, through several domains), give it the value of Balmer's number. Let us look at the recessive and expansion processes in the light of this.

In the recessive process, something arrives from the next-higher domain, e, which, in turn, received it from higher ones again. This distorts the equilibrium "by adding new blocks." We cannot speak of "mass," "energy," or "velocity" but only of mutual action of the recently arrived particle which rearranges the system. Then what happens? There is no lower domain into which this tension may overflow, as there is in the case of the higher domains. There is but one escape, and it is toward e.

174

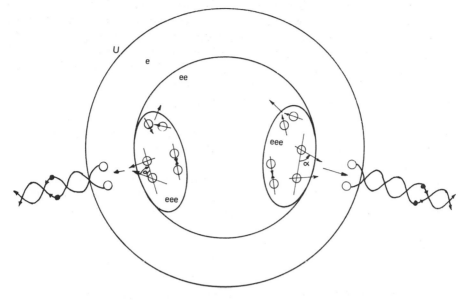

Figure 50

When a certain level of tension of the already tensioned state is exceeded, there is an expulsion.

Up to this point this was the recessive process, and this is why, in U, we believe that all masses converge within the process of rearrangement or distribution of available energy in available mass, compressing matter and giving rise to an ever-greater tension, which, on reaching its limit, produces an explosion, thus becoming the expansion process.

In the expansion process, in eee, since the particles are in equilibrium—we suppose in pairs—then the field distortion, as elemental mutual action between them, produces a vector-angle change of vectors originally equalizing each other along the same line of action. Under these conditions, a binary pair is ejected. This constitutes the initial phase of an elemental mass organization in ee, having a certain velocity that we call unitary.

From this expansion, this energy transfer, from this battery, the eee domain, where the energy had accumulated in the foregoing recessive phase, it is possible to follow the process and all the steps that give rise to detection in U, by ourselves or via instruments, of things we call "radiation," "heat," "electromagnetism," et cetera. All these phenomena have the same physical basis, even though we detect, for instance, heat as molecular agitation. We should look more closely to see what this agitation is.

If the tensioned state gives rise to radiation—energy—which in its gravitational manifestation we call supergravity, we recollect that we have electrical discharges in a gas emitting radiation from lower domains, and we forget for a moment that the third postulate appears to be violated (where does the electrical energy come from?); then we see that we have performed some sort of recessive process. If this were the case, we could generate supergravity in the laboratory by progressively tensioning a mass somehow (not violently—e.g., electrically) such that in eee the expulsion threshold is not exceeded but rather our sample remains as a charged battery. Under these conditions, could we gain energy by obtaining more than that provided?

If this were the case, could we gain it using any element or combination of elements? Will there sometime be an artificial uranium? These are questions that will only be answered sometime in the future. The replies may well have some practical bearing in the area of future energy supply.

An interesting question would be that of whether clean (radiation-free) atomic explosives may be obtained. Most of the damage from a fast fission or fusion reaction is caused by radiation, hence a phenomenon from domains below e but reaching us in U. This clean type of explosive may be difficult to obtain, but if it were possible we would have only mechanical effects such as with TNT. The advantage would only be in the comparatively tiny amount of material necessary. A further advantage would be the controlled use of the energy available independent of uranium.

Under these conditions, there would be no risk of explosions at atomic plants or of damaging radiation.

In order to reach this goal, two things are needed—namely, that it be shown that the energy balance is favorable using normal materials and the development of a process to attain a controllable state of tension in the lower domains.

176

Chapter XIX

INTERFERENCE OF POTENTIAL FIELDS

INTERFERENCE IN THE U, e AND ee DOMAINS

We had asserted that in local regions of the universe (U), there is a self-evident tendency for momentum to pass from zones of high-field resultant to those of low-field resultant.

It is not difficult to understand the mechanics of such an energy transfer or passage in one single domain. The moment we want to visualize the passage between domains, matters become so complicated that our formulae, with their constants, and a certain shortfall in our comprehension of phenomena, lead us to a somewhat confused idea of the phenomenon. Thus we should find an equality between both phenomena (in a single domain and across a domain boundary). In this line we have already included the principles of thermodynamics, electricity, *et cetera*, in a single conceptual unit. We now need to exemplify the passage from a quantifiable mechanistic standpoint so as to bring the phenomenon from e into U where we will classify it appropriately with other similar phenomena with which we are familiar. In this way, we can accept the equality.

To do this, after finding the interference resultant of the potential fields created by the masses in each domain, we shall make the appropriate scale change.

First of all, we shall consider that in the void of any domain, there appears "a" mass. It has a latent field manifested by the presence of another mass, and this is a reciprocal phenomenon. Since the field extends to infinity, we shall not consider distances. Also, this is theoretical and may only have occurred once or twice in the universe (in any domain). Once under these conditions, we can speak of assigning a network of equipotentials and lines of force. We may set up the network arbitrarily, giving, for example, a maximum value at the center and a value of zero at infinity. We will select this convention, although we could have chosen the inverse one.

For the spacing of the network we have two options: equal or unequal spacing. If the spacing is equal, the field values corresponding to each is a different fraction of the whole, while if the spacing is unequal, each can be such that the corresponding field values are equal fractions of the total. We will opt for this latter approach.

It is hardly necessary to mention that the equipotentials and the lines of force must cross one another at right angles, and, for the flux calculation, they must, as far as possible, form squares.

177

In Appendix A we will see the following three cases succinctly:

(1) simplification in a plane;
(2) the case of three bodies in a plane;
(3) the case of three bodies in space.

These cases are, as far as we know, valid for U, e, and ee. We shall assume that in e, potential field interference produced by e domain masses does not give rise to what in U we detect as waves, but that waves are produced by ee domain masses moving through the potential fields of the e domain. The reason for this is that, apparently, field interference in e produced by e domain masses produces heat, electricity, *et cetera*, because the masses are too large for their own domain. On the other hand, interferences from ee passing to higher domains produces a distortion such that in U they are waves. For this reason, it is interesting to note that, in view of the considerations to be seen in the next chapter, light is incapable of leaving U and crossing into UU. This would lead us to think that the light of UU will be something arriving in that domain from its lower domains. We don't know what it is nor, indeed, whether it exists, or if anything arrives in UU; but if nothing arrives there, there will be no equivalent of what we, in U, know as light.

QUANTIFICATION OF THE PASSING ENERGY

In order to do this, we must start from general principles. Once these have been brought to mind, then to know and visualize from U the passing of energy we have to face the following difficulties:

(1) Between the domains we are considering, masses have different structures.
(2) Velocities are equal.
(3) Times and distances are "comparatively" different.

The passing of energy is, therefore, quantified as v^j, the mass, as what it is in each domain, as m_j. Then, in order to establish a visualizable comparison, we change the time or distance scale (but not both at the same time), hence distorting the e domain phenomenon and introducing it into the U domain, thus to visualize how much faster it is with respect to U.

If we take a velocity as unitary with respect to the local time of e and the distance covered with respect to the local time of U, it will be very slow.

For example,

If $Ve = 2.3 \times 10^{10}$ cm/sec
and $T/t = L/l = 6 \times 10^{21}$
then $l = 2.3 \times 6 \times 10^{10} \times 10^{21}$ cm $= 1.4 \times 10^{32}$ cm.

That is to say, in that time unit (1 second in U) the distance covered will be far in excess of that corresponding to c (3×10^{10} cm), which we take as the maximum possible. Also, the value 1, as a product of the distortion, is impossible.

This calculation illustrates to us that time is subject to a domain passage constant on passing from one domain to another.

COMPARISON WITH QUANTUM MECHANICS

From points (1) and (2) above, we can see that there must be more determination using this method. This is to say, by appropriate application of time and/or distance scale changes, we can transfer e domain phenomena into the U domain so as to consider them in more detail.

Once conclusions have been arrived at in U, the results may be transferred back into e, using the inverse of the scale changes employed originally.

This is analogous to the Cartesian method of analysis, in which a physical problem is transformed into a mathematical proposition. The work done is exclusively mathematical, and the results are reapplied to the physical situation in order to facilitate interpretation of their real meaning.

179

Chapter XX

COMMENTS ON RELATIVITIES

SOMETHING ABOUT SPATIAL RELATIVITY

We shall consider here what seems to be the enunciation that space is relative. This is the same thing that happened to time in the theory, special or general, of relativity, and we are led to it by circumstances. We are familiar with the history of the theory of relativity and know that it is a real, mathematical, and justified fact.

We now find ourselves studying a complex phenomenon that may be summarized in the phrase "space is relative." This will be of great assistance since with it we can express the complexity of what we have been laboriously explaining. It is also a real, mathematical, and justified fact.

It arises from the parallel worlds, each with its own space, each belonging to a total or absolute space.

In the relativity of time, time itself was not absolute. In the relativity of space, space itself is not absolute.

In terms of relative time, phenomena are not the same for all observers, and, likewise, in terms of relative space, phenomena are not the same for all observers.

When time is relative, phenomena in which it is involved are not absolute.

When space is relative, phenomena taking place in it are not absolute.

The quantification of phenomena in relative time has already been amply developed and understood.

The quantification of phenomena with relative space is what we are attempting to develop in this essay.

Even so, these efforts lead us more toward the unified field than a physics of relative time and space. This, in fact, becomes a necessity if we are to arrive at a description of reality.

In other words, today, physically described reality requires both relative time and relative space.

If time has been called the fourth dimension, space may well be the fifth, except that in the term *reality* we encapsulate only what happens in one domain.

When we use the term *space*, we refer to that space in which masses are located. It must not be confused with space as relative distance, resulting from the application of relativistic formulae, or with the three dimensions defining it in a domain.

Relative space arises as a parameter when we wish to analyze reality in terms of the structure of matter. This means that, basing ourselves on the phenomena of U and e—the atomic and universal levels, as definite realities—we analyze physical phenomena as components of that reality, making full use of the mathematical relationships between them.

Upon so doing, we see that when space "contracts" (This may not be the best word), similar phenomena, measured from another domain, occur more rapidly. Even so, since the distance, time, and velocity scales must be the same in U and e, similar phenomena in those parallel worlds are such, with the appropriate change of scale.

The value of the scale change quantifies relative space between the two domains considered.

The value of the scale change tells us how a phenomenon must change in the space of one domain such that, upon transfer to the space of the other domain, that phenomenon will be the same.

If these spaces were not relative, but there were but a single absolute space, a phenomenon transferred from one domain to another without a scale change would not necessarily remain the same. For example, we may have the following:

(1) Space is absolute.
(2) We transfer the phenomenon "electron rotation about a proton" from the e domain to the U domain.
(3) The phenomenon ceases to be the same since, in order to maintain the similarity, its speed in U will be that of the e domain multiplied by a scale change greater than one. This is in conflict with the conservation principles since the applicable value must be one.

REGARDING THE RELATIVITY OF TIME

There is little to add to this relativity; it was an obvious intellectual achievement. It does, however, lead us to think that, given the structure of matter, we can speak of a relativization of space. This, though, is not the place for labeling matters, but rather for clarifying them.

The quality of time and distance scale changes in the passage between domains indicated the justification of applying the relativistic formulae in each domain separately, with the criteria we stated in their passage between them.

If this were not done, the explanations of phenomena to which we do not have direct access would become complicated. The possibilities of complication are real, we have only to take the case of the second

181

Postulate: "The velocity of light is independent of that of the luminous source."

In this I want to adopt a personal position. I had always accepted it, and it was necessary to postulate it. Since I already had a mental idea of the structure of mass, it was no great difficulty to accept that once light leaves its domain, it is independent of its source, or point of origin.

We now get to something a little confusing, namely, the red shift. We have all been told of the shift toward the red end of the spectrum and that it is the result of the "influence" of the source with respect to the direction of departure from it of the light concerned. How can this come about? Something, obviously, is wrong.

Let us review the phenomenon. In the wake of the tremendous expansion of railways in the period from 1850 to 1950, almost everyone has experienced the rise and/or fall in the pitch of a locomotive's whistle, depending on whether it is approaching or leaving the listener.

During the same period, astronomy also made great advances, and it was natural enough that this (Doppler) effect should be applied to light waves, so explaining the shift toward the red as evidence of the stars in question receding from the observer (ourselves) at this or that velocity.

Is this, in fact, the case? If it is not entirely true, should we accept it for the time being? I think not.

Let us consider the facts. In its movement, the sound source "compresses," as it were, the waves emitted forward along the line of its direction of movement, advancing, as it is, through a compressible medium, the air. In the same medium, the waves emitted backward along its line of movement are lengthened.

There is a medium in which everything concerned takes place. We are familiar with the characteristics intrinsic to it, and those of how the perturbation is caused. In other words, this is a U domain phenomenon having implications probably no further afield than the e level.

Is light in some way similar? Is the medium involved in the case of light similar? No one can really say what it is with any certainty.

Let us try out our field solution. Light has a supposedly constant speed. We have already noted that if it is stopped or absorbed, its speed—or, at least, some of its energy—can diminish. If this were so, what would we see at the e or ee level? We can look back to Figure 44. There we see that the rays of higher energy and frequency suffer a lesser refraction, or deviation, than the others, and this is logical. The lower energy rays are refracted more, moving toward the higher wavelength end (the red) of the spectrum. We assume the same field values for all the atoms of the crystal lattice.

What, then, happens in the case of the light from distant stars, and why is there a red shift? Could it not be that the light—having crossed vast regions of space and having had its own potential field interact with the innumerable other fields of sidereal masses, gases, atmospheres, the enormous fields of other stars, *et cetera*, similar in quantity to the atoms of the crystal lattice—suffers the same effect as in the case of the prism?

If this is correct—that is, the light arrives at the earth with its energy somewhat attenuated—the relative speed of the star would be quite unrelated to the effect to which the light was subject. The energy loss would be due to a reduction in the rotation ($\overline{v}$) of the photon (whatever may be the photon structure adopted) and not some effect on $\overline{V}$, the translation velocity equal to c.

In this case, the velocity of the source could only be arrived at by using the difference in time of two red shifts, for which we would assume the energy reduction of the light is linear and directly proportional to the amount of space covered between source and observer. The assumption is approximate, since two rays may pass through space and encounter different sets of potential fields. If, however, we take two rays passing through the same chunk of space (the laboratory being in the same place for a second reading, one year after the first, and neglecting effects of the sun), we could arrive at a value for the velocity of the source.

The conclusions from all this would be as follows:

(1) The expansion of the U domain as arrived at on the basis of the Doppler effect must be revised.
(2) If the velocity values of distant stars and galaxies are corrected, we would have other values for the expansion process of U.
(3) The lack of energy reduction in the light of "nearby" stars would be due not to their low velocities but to our not being able to detect an appreciable reduction, even though one exists. (This last has already been shown for light from the sun). By the Third Postulate, light also tends to transfer velocity to higher domains, otherwise the postulate would not be complied with.
(4) It is worthwhile mentioning that if all the foregoing is the case, there must be a source sufficiently distant for its light to die out altogether before arriving at the earth. This would mean that the limits of the universe given by light lie at the largest distance from whence it is possible for light to arrive at all. The real limits must be more distant again, but light from those regions would have gone beyond the red into the region of radio waves. Could this account for some of the signals detected by radiotelescopes?

The implications of all this are, of course, something of a departure from current trends. It would mean that the limit or boundary of U, beyond which lies UU, would not depend on a light signal but be beyond the distance from whence light may arrive at all.

We have thus made some progress regarding the limits of the U domain.

If we were to place ourselves far enough away from the earth in any direction, then, with the same range of vision or detection, we would discover new "distant" bodies in space—beginning with their spectra creeping in at the red end of the spectrum.

If, on the other hand, we were to stay here and develop better methods of infrared imaging, we would see objects more distant than we do at present since we would be detecting light that, with its diminished energy, barely made it here at all.

This, of course, should not be confused with the improved infrared images achieved by a European satellite focusing on visible bodies.

While it seems to me highly unlikely that the Doppler effect for light will be readily discarded on the basis of these arguments, they do appear to be well founded.

If (1), (2), and (3) above were not true, then

(a) it would be impossible to look at the sky because, with the light from all celestial bodies arriving, it would be too bright.

(b) the limit of U would be that of the "range" of our telescopes.

SOMETHING ON COSMOGENY

Since we touched on the matter of the Doppler effect, I would like to comment on the apparent unanimous outward flight of distant galaxies associated with it.

By Hubble's law, the recession speeds of heavenly bodies are proportional to their distance from us:

$$V = H.D = 260.D,$$

where D is the distance in megaparsecs and V the velocity in km/sec.

In other words, the greater the distance, the greater the speed of recession. This would be in keeping with the considerations of the "energy loss with distance" interpretation of the red shift being a possibility other than a Doppler effect.

Consequences of this would be the necessity of changes to Hubble's expression, and the speed of recession would not necessarily depend on the distance of the particular body from the earth. This last may have more to do with celestial mechanics, whereby several galaxies constitute an even larger mass, the parts of which may have different rotational and translational mechanical values.

With the above expression, the greater the distance of bodies from the earth, the greater their speeds of recession—and these speeds can be as high as ten percent of the speed of light.

Once again, we find ourselves at the center of the universe, or, rather, that our galaxy is at the center. Why should everything be fleeing from the location of our galaxy? I wouldn't be at all surprised if those distant galaxies had a behavior similar to that of our own.

One way of expressing the speed could be as follows:

$$V = (Å_{t_2} - Å_{t_1}).K,$$

where Å is the wavelength of a particular line of the spectrum, t_1 and t_2 are the particular times, and K is a constant for the falling off of light energy per unit distance.

If we take the sun and the sodium lines, we would observe that the shift is undetectable, the line remaining at 5,890 Å, while for a distant nebula the same line shows up at 6,480 Å.

The value for K may be estimated by dividing the difference between these two values by the distance, D, to the distant nebula as determined by another method. Then:

$$K = \frac{6480 - 5890}{115 \text{ meg.}} = 5 \text{ Å/megaparsec.}$$

(1 megaparsec = 3.086×10^{19} km.)

$$D' = K (Å_{t_2} - Å_{t_1}) = K Å_{t_2} - K Å_{t_1} \text{ (Km)}$$

and

$$V = \frac{D'}{t_2 - t_1} \text{ (km/sec.)} \qquad\qquad \text{(XX-1)}$$

The $(t_2 - t_1)$ value must be as large as possible, since if it were small the $(Å_{t_2} - Å_{t_1})$ term would not be discernible and we would have the same situation as occurs with the sun—namely, that of a lower velocity and an undetectable shift.

The upshot of all this is that for a spectrum line shift of one Angstrom unit, there must be a distance of 3.086×10^{19} km covered and, for light, this corresponds to a "trip time" of three million years.

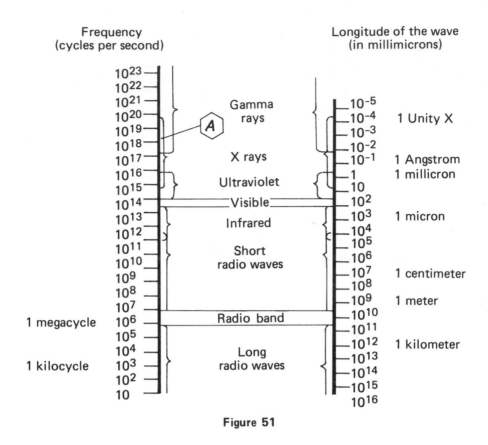

Figure 51

Figure 51 shows a wave scale in which we can see that if the shift is pronounced, then using that scale alone, we could determine the distance of the origin of any particular light the wavelength of which has become progressively longer, possibly even reaching radio-wave frequencies. Thus we would have a first connection between light—generally seen as an electromagnetic wave—and what is outright electromagnetism—namely, radio waves.

To further analyze this transition goes beyond the scope of this essay, but I think that the consideration of the unified field formulae in Appendix B may be of assistance in understanding the phenomenon.

The diagram below is a representation of wavelengths in millimicrons, from 1 to 10^{16} millimicrons. The value X would be the maximum distance from which we could obtain information—in other words, the limits of the universe U accessible to us.

186

$(10^6 \text{ milli } \mu)$ 1mm

10^{17} Å

$(10^{16} \text{ milli } \mu)$ 10^4 Km

$$X = \frac{10^{17} \text{ Å}}{5 \text{ Å/Megp.}} = 0.2 \times 10^{17} \text{ Megp.}$$

$$= 0,62 \times 10^{36} \text{ Km.}$$

$$= 6 \times 10^{21} \text{ light years}$$

The light year value of 6×10^{21} is the same as that of the U/e scale change for distance or time and is the second—a dimensionless value.

The immediate question, of course, is that of whether this is a simple coincidence or does it have some physical significance.

If we want to make an analysis using the electromagnetic wave scale in Figure 51, we could arrive at the following conclusion: Phenomena being compared should be described with the appropriate parameters—where they actually occur—using the laws that govern them, *et cetera*, so as subsequently to determine the possible connection or sequence between them.

From the A section of the brackets in Figure 51 it would appear that the same frequency/wavelength can have different origins—*i.e.*, the transmission of field perturbations arising from different phenomena. Thus, for example, gamma rays would not have the same origin as X rays, although the perturbations (the waves produced) could be the same, and neither type would become the other under any sequence. In other words, frequencies and wavelengths would be independent of the phenomenon producing them; this would support the concept of the wave therewith associated being a field perturbation.

It seems that all such waves originate in the ee domain.

We may well ask ourselves, what are simple radio waves from broadcasting stations? What are, for instance, the short waves and where do they come from? A hurried answer would be that they are from the ee domain: This may be possible, since light is formed in the ee domain, and we saw how closely related light and radio waves are. Everything; therefore, seems to point to the ee domain as the point of departure of radio waves.

Without going into any further detail regarding ee domain phenomena, they seem to have a great breadth of manifestations, though not necessarily complexity. This lower domain group of phenomena is perhaps simpler, though the manifestations are many. The e level phenomena appear to be more complex, and there is somehow a beauty in imagining waves effortlessly passing through domains.

187

We should note that if radio waves are received from some particular point in space, it does not mean that these necessarily originate in some star located there, though, should there be one, and the waves do come from it, why should it not emit both light and radio waves? Indeed, why shouldn't an emission, starting off in the infrared, finish up being received as a radio wave? All these cases may occur, and possibly many others.

In Appendix C there is a deductive exercise on why, with an increase in wavelength, there is a reduction in energy, when at first blush the opposite would seem to be the case, corresponding, as it does, to a more distant orbit.

REGARDING BIOLOGY

As a point of departure, we may use the following definition of life: "Life in each domain is the most complex form in which matter can organize itself, given the possibilities of combination available to it, with the elements from lower domains."

It has been said that there is something of magic in science and one must always be prepared for the unexpected. When we arrive at some concepts and are surprised at the result, there is something like a magical effect; we are momentarily paralyzed. Our reflex reaction is one of pleasure if we find something of truth in our results if, for instance, we realize that light illuminates and discover it to be so and that it was the mystery that had puzzled us.

In living matter, the first thing we should look at is the effect of the actions attributed to the supposedly existing beings in it.

If we can find out something of these effects, we will be in a position to infer what is producing them and then even something regarding the physical characteristics of the beings.

In this, there are three stages which may usefully fall under the following headings:

(1)   the effects of the actions of the beings;
(2)   what may produce such actions; and
(3)   the characteristics of such a "producer."

Of these items, the first is the one we can most readily establish, and establish completely. The second point leads us to a series of suppositions which will be greatly influenced by modern knowledge of electronics; and the third point may become an object for personal speculation. In bygone eras, this freedom for the imagination gave birth to numerous sea monsters when man was unable to cross oceans or was apprehensive in regard to doing so. Let us hope we can maintain some reasonable bounds on imagination this time around.

(1) *The effect of the beings' actions has to be such that these individuals of the microcosmos produce, in U, what we know as animate life.*

Since it would seem that matter becomes simpler as we pass downward to lower domains, we should locate our beings in the e domain. This is so insofar as they themselves must have some degree of compatibility, which would only be available if the prevailing structure of matter can rest upon two lower domains before reaching the level where the border between matter, energy, force, and velocities becomes blurred.

We have to assume that our beings reside in organic matter. This, of course, means, at the e level, a fairly complex level of organization of matter. The question of whether such beings may be present at the e level in inorganic matter, while interesting, is irrelevant for the moment.

As was already mentioned in "The Unified Field," the effects of the beings' activity as observed at the U level are thoughts, imagination, instinct as a particular case of thought, *et cetera*, all of which were intended for sustaining life.

If we think of ourselves as embedded in nature, try to describe ourselves as a consequence of its dynamics, and that that which determines what we are as living beings originates in that microcosmos, we arrive at the conclusion that our free will is not so much "will," nor is it as "free" as we may have liked to think. This would support the thesis of those who assert, with a certain fatalism, that our destiny is predestined. This may not be so fully the case, but neither are we entirely free. We are bound by laws of mandatory compliance and are as much a part of nature as an atom, planet, or a galaxy, formed as we are from the same elements as they. Even so, we are still interested in what it is that facilitates our being different from inorganic matter.

This is why we went through the two stages from inorganic matter— first to the organic and then to the appearance therein of a something having very particular characteristics, the effects of which at the U level we have already determined.

(2) *The actions spoken of here could be produced by entities with certain restricted vital characteristics making the beings simpler than ourselves.* This would be due to their being in e and having their matter organized as it is in ee. In other words, their bodies would be formed with the mass having ee characteristics, though

everything must be simple except the intellectual results. If we put it another way, we would say that beings are of a simple organic constitution but their intellectual capacity for carrying on in their domain must be as good as ours for doing so in our domain. If we assume they live only in complex molecules, then, if we were to shrink down to their size, their habitat would appear quite complex. The reason is that while we have assumed a lower level of complexity in the organization of matter in that domain with respect to U, we don't know how much less complex it may be.

Even so, we can hardly assume that they are less intelligent with respect to their environment than we are with respect to ours. If such were the case, we would encounter the problem of generating something greater (ourselves) from something lesser—for example, some type of animal.

Could it be that the superior being in the U domain (ourselves) could have an e domain basis in a grouping of beings individually inferior?

This may be possible, as some manner of analogy is the case of the computer's "current on or not," but although the machine is producing myriad marvels, it is not producing intelligence. We may ask ourselves what, in U, is intelligence. Having seen that it is a manifestation of the e level beings, we have that that manifestitation is a component or part of intelligence. The beings, therefore, must have intelligence. The conclusion would appear to be that their intelligence, while not of the electronic type at their level (e), is similar to it at the ee level—namely, the existence or nonexistence of some force. At their level, e, in the "thinking matter," this gives rise to a first spark of intelligence. The union of all the beings is manifest at our level, U, as our own intelligence.

In this way the principle of complexity would be maintained; the equivalent to the electronic process is located precisely in that domain where matter takes on the simplicity of elemental forces and mutual interaction. In this way, starting off from the eee domain, we can arrive at U, describing the organization of the most complex form of matter of which we are aware, which, in U, gives rise to the human being, ourselves. It would appear a little complicated to try quantifying this phenomena through the domains using passage constants (scale changes).

(3) *The physical characteristics of these beings are, at best, difficult to imagine.* This is because the beings, if they are formed by mass organized at the ee level, as equivalents of our atoms and molcules, would have photon-type particles and others only now being discovered (see particles listing in "Tables"). The relative

abundance of such particles is a good sign insofar as it would support their structure's being relatively complex, though less so than our own. This would also fit in with their intelligence being based on the on-off computer system where the beings would, by a selection process, make it more sophisticated using only particular solutions arising from their own inner selves (remember the Russian doll syndrome?) or even simply using some mixture approach such as the idiot's method. At that level of simplicity the idiot's method may even be very efficient, given the paucity of information permutations possible. We ourselves can hardly use it in view of the number of variables involved. Even so, we don't really need it, since we do have intelligence, and where can it come from but the lower domains where the idiot's method may be used. Thus could mass, matter, and life be organized.

If we see things so, we may compare the two computers—man and that constructed by him—to see what differences there may be. We would then see the following:

(a) The computer or processor produces only those results that are in keeping with a previously planned sequence.

(b) As far as we are aware, these results have no basis in living matter, with beings in a lower domain producing them, but rather that, at that other level, we manipulate the forces such as to obtain the desired results.

(c) If we were to assert that results may be the same using either method (*e.g.*, sums totalled by a person or a computer), we can establish the difference not only by origin but also by using the following comparison: transport by a person or a motorcar. The one is rational and the other mechanical, but the result is the same. We may go on to say that the processor is not aware of itself, since (i) if it were to have self-awareness, it would have acquired it via processing, hence it would be something other than that of man; and (ii) if it were to have a self-awareness equivalent to that of living matter, it would have acquired it via the lower domain beings, thereby falling not into the category of "artificial intelligence" but that of "artificial life."

We can summarize by saying that, in the case of the processor, computational phenomena are produced having an underlying artificial logic but no rational logic. The first of these gives rise to "artificial intelligence" while the second, being based on an entirely different phenomenon, produces not only artificial but also rational logic.

191

Basically the difference is that, in the processor, processes in e are manipulated from U, while our intelligence is the reverse inasmuch as it is the consequence perceived in the U domain of events (the processing of which we are unaware) having their origin in the ee domain.

CULTURAL IMPACT

It is worthwhile to make some observations here in regard to the cultural significance of these new proposals. In "The Unified Field" as well as in this essay there are a series of conclusions of a nature such that they ought also to be considered in the cultural context.

By cultural context I mean rather the knowledge and use to which they may be put among the population at large, to include therewith those of a uniform level of culture and folk of higher achievements in areas other than those treated here.

For those of us whose lot has been to work for years with potential fields it is no problem to accept the study of forces in them; it can be appreciated how much more elegant (so they say) it is to forget about forces of attraction and conceive of matters as the interactions of potential fields and so better explain the gravitational phenomenon, its detailed calculation, and its complete rational acceptability.

For folk in general, however, when will it be accepted that bodies are not attracted by the earth? That they move freely through space and approach or separate from other bodies depending on the amount of velocity they carry? That this applies to all bodies and not only the (more readily acceptable) case of heavenly bodies? That bodies falling to the earth are responsible for their fall? That had they been carrying more velocity they would have moved on elsewhere?

I can but suppose that numerous explanations will be thought of to convey this concept, but only when it is widely accepted can it be said to have achieved some cultural impact. Even so, that is hardly the end of the story.

Undoubtedly, the next point on the dissemination agenda will be that of a straightforward explanation of the structure of matter. Today, two domains—the universe U and the atomic e—are generally accepted and familiar to many but how should one approach the explaining of further domains that were necessary for our investigations of some other phenomena and are now accessible to us, mostly through our instruments? The Russian doll approach is good, I think, but there must be other better illustrations. The number of domains presented here is really only accessible to the imagination, and unless evidence appears to support them, their acceptability, even as a theory, will begin to evaporate.

The cultural impact of having described a perfect machine in the order of the cosmos may give rise to some religious fervor, a surprise of some kind or other, but one way or another, it would affect those becoming apprised of it.

The simple fact that, in order to explain the overall phenomenology, we were constrained to postulate some principles above and beyond those of classical mechanics would indicate that some cultural significance should be thereto attached.

One asks how people may receive or interpret the fact of energy coming from the interior of masses when it is plain and obvious that when I lift a weight, I lift it with my own muscle power.

In time, of course, everything will fall, as it were, into place, and it will be understood that in our expanding universe (a simple and generally accepted concept) each mass must have had some thrust from the original Big Bang. How, though, should it be put across that such "thrust" is being exchanged and that it arises from the interior of the mass itself? This is an awkward task, at best. We can only hope that better explanations than those here presented will come with time, so as to enable a clear picture to be reasonably presented.

A further aspect of the problem in the general culture context is that of textbooks. The explanations here are so far removed from the traditional ones as to challenge the intellectual honesty of anyone meeting them. Those interested should at least compare them with the traditional approaches with a view to arriving at their own conclusions. These, in the final analysis, are the only ones most folk will accept—in itself a laudable point, since, in this way, new contributions can be generated.

Thus the textbook matter could give rise to a transitional period, during which conceptual structures would undergo something of a remodeling until new improvements come along. This sort of thing takes years, but the cultural shift at the student level is a necessary adjunct to the overall conceptual reshuffle.

The Fourth (unwritten) Postulate is that of a certain similarity of phenomena from domain to domain. This would entail at least provisional acceptance of the assumption for explaining the manifest properties of living matter: that at the molecular or possibly the atomic level, there is a something we have called *beings*, having a certain similarity with what we observe ourselves in the next higher domain, our universe.

This is something so incredible even as a postulate that all we could really say is that these beings resemble ourselves only in their manner of manifesting themselves, since our minds can hardly be asked to conceive of such large scale changes (10^{20} or more). With such a postulate, we would be in line with the principles (postulate accepted but not demonstrated) of phenomenological similarity from domain to domain.

How about cultural impact in the above matter? The first logical reaction is outright rejection since it looks as though we have taken a problem and conveniently transferred it to some mysterious region whence, in due time, it may well be shoved further onward to some new one. This, however, is hardly giving the matter a fair hearing because, in the "down" direction, it would appear that there are only three steps—from U to the eee domain to the end of the road. Moreover, on the way, the complexity of phenomena diminishes progressively until matter and energy become blurred.

There is a prevailing tendency to see Martians or little green men at every turn, but, even so, if we are interested in what goes on down there and how, in the U domain, matter has a will, we cannot circumnavigate this business with olympian nonchalance. There *is* something there, call them what we will, whether little people, beings, chips, or bits.

As for the cultural influence in folk at large, this may engender speculation and certain flights of imagination. In this area, biologists and psychologists can help by assembling coherent theories in their respective disciplines, based on the unified field. This may curb the more hairy speculation and the proliferation of low-grade science fiction.

It is my hope that this essay will provoke more questions than the answers proposed, thus to further maintain a healthy imbalance between the two. If we ever get to the point where the numbers of questions and answers are the same, the ascent of man would be at an end. As questions were answered, others not being forthcoming, the descent of man would be getting underway, to continue to a final universal state of repose and oblivion—if such could really exist.

This is as far as this train goes. The card game with the Supreme Creator continues. He gives us any hand we fancy, but it seems we can only identify two cards—namely, matter and movement. The game, while not poker, is not dissimilar, and it's our deal; but what is the next move?

An audience watches the stage as the actors prepare whatever it is they have in mind, and He smiles benignly.

In the unification process it looks as though the next stage should be that of giving a consistent logical underpinning to Figure 51. This wavelength scale gives a number of apparently different phenomena and we are told that they are different manifestations of the same one—namely, the electromagnetic phenomenon. This itself, however, has yet to be adequately explained.

In Appendix B there may just be a key to this. During our considerations of the Doppler effect, we were able to find a transformation from light to radio waves, all part of a single phenomenon. In the same way we could, in the light of what we know of the structure of mass and the principles here proposed, go through the scale looking into characteristics of phenomena giving rise to different frequencies/wavelengths to see whether there is a relationship between them: Are they really different, or are they the same phenomena that, in one way or another, have been transformed such that they may be seen in two guises.

This would be a new intellectual adventure covering many pages and which could probably be condensed into a few formulae. I chose another route, however; the formulae could be those of Appendix B, but these are cold and say nothing.

I have left the pleasure of wandering down new avenues, exploring new paths, and discovering new and breathtaking landscapes to the reader, thus to promote reflection and discussion between colleagues and the benefits thereof.

Appendices

To summarize, we have come to the end of a theoretical study in which, with the healthy intent of improving knowledge, many experimentally unsupported claims were made. The only experiments cited were those already generally known; this essay, however, suggests a great many more and these experiments will, so to speak, have the last word.

In view of the above, all shown here is tentative and, at most, explains the structure of matter in a more advanced way and interprets the dynamics we observe around us in an effort to give it some overall unity and sense.

To refute something properly must not mean to replace it with the contraries but better to unify things—in this case, different areas of physics—using proper judgment.

If this should come about, then this attempt will not have been in vain.

Let us finally underline just a few of the main points made:

(1) With the practice gained by having read this far, the reader will be locating the unfolding of phenomena in the today unchallenged structure of matter.

(2) We have found in the energy-domain passage rules a first-order relationship justifying the higher orders of velocity and absolute temperature.

(3) We have managed to put together an improved enunciation of the law of gravitation.

(4) We have been able to enumerate three postulates more general than those of Newtonian mechanics, which is encompassed, to-together with those of thermodynamics. These are as follows:

(a) *The Mass Principle*: Mass has value but hardly exists. The aggregate of such values of mass is a universal constant.

(b) *Force Principle*: The total of forces is a universal constant.
Corollary (i): All action is always opposed by a reaction and the mutual actions are equal and opposite.
Corollary (ii): Energy cannot be destroyed.

(c) *The Conservation Principle*: Apart from the state of absolute repose, forces tend to distribute themselves among masses from lower domains to upper domains in the expansion phase of the universe and vice versa in the contraction phase.
Corollary (i): All bodies remain in their state of rest or uni-

form motion unless acted upon by forces exercised upon them.

Corollary (ii): The change in movement is proportional to the force applied and in the direction thereof. Momentum is proportional to mass and velocity together.

Corollary (iii): Entropy cannot be destroyed but can be created.

We have also generated an estimative table of constants of total field resultants for all domains and of relationships between different indicators of such resultants.

We have found a material explanation of the Doppler effect (in astronomy) which may be of some significance. We have looked into the standard kilogram and put together a new set of equations for the unified field (Appendix B). Even so, rather more important is the conceptual description of the structure of matter, which allowed of all this, and should lead progressively to a better understanding of nature, a part of which we are and in the dynamics of which we take part.

The future appears promising and full of pleasant surprises.

Appendix A

MORE ABOUT
POTENTIAL FIELD INTERFERENCE

As an immediate consequence of Gauss's theorem, we find that the field generated by a homogeneous spherical mass, or one of concentric, constant density layers, is the same outside the sphere as it would be were the mass concentrated at the center. The potential field interference calculation in its simplest form is as follows.

We take a body—for instance, spherical (Figure 52)—and we cut it through a plane, alpha. We assign a field value, VM, to the center O, where V is the sum of the rotational, translational, and any other velocities (*e.g.*, a vibrational term), and M is the mass.

If, at point O, VM is equal to 100, then at infinity, VM is equal to zero. From the point O to infinity, we can put in adequately separated equipotential lines.

If at each point on an equipotential, we place a test mass S (as in Figure 52), the field value T manifested is given by T = 40 – 30 = 10. The test mass S, however, also has its own field and the interference between the two determines the resultant. The resultant's direction sign,

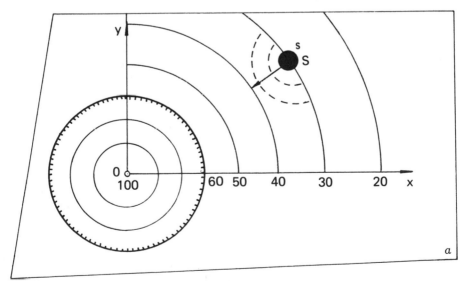

Figure 52

201

and value will be the difference between the two vectors appearing as a result of the presence of those masses, with their values and velocities, at those points in space. This is the case for all domains.

In Figure 53 we have two bodies, one test charge S, and there are three interferences, but we do not consider that of S.

Now, let us look at the three-dimensional case (Figure 54). This we present only in its diagramatic form, since in this essay we have always avoided mathematical expressions and particularly spherical coordinate representations of potential field interference, which take up a great deal of space.

In Figure 53 for similar reasons, we have made no differential analysis of the S – S''' trajectory.

We observe here that, although only one trajectory is drawn, there is a series of actions continually varying according to the position of the body in the field interference pattern. The resultant at each point is fed, or increased, by the force of the gradient at that point—in other words, a velocity which because of the nonuniformity of the field pro-

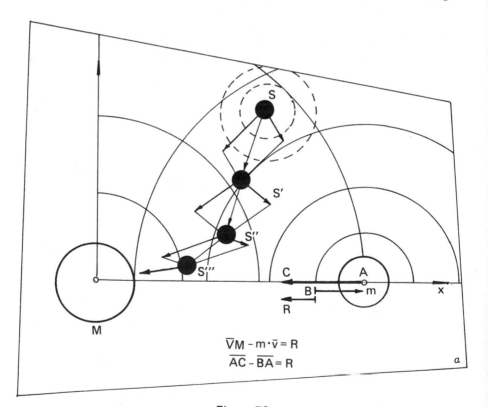

Figure 53

202

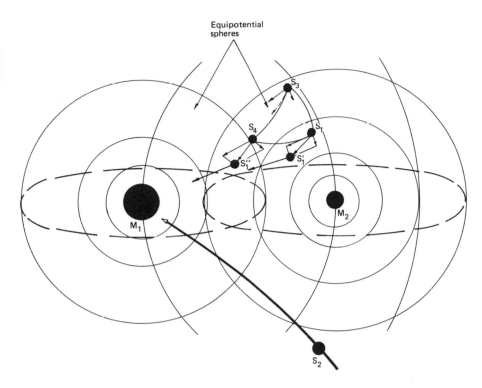

S_1, S_1', S_1'' ————————— trajectory of S_1 toward M_1

Figure 54

duces a variable resultant velocity which because of the nonuniformity of the field produces a variable resultant velocity—*i.e.*, an acceleration. In the eee domain, all this is simplified down to two bodies and a single mutual action.

We should consider the following cases:

(a) Simplification in a plane;
(b) The case of three bodies in a plane;
(c) The case of three bodies in space.

(a) SIMPLIFICATION IN A PLANE

We simplify the case of a body giving rise to a field by taking a plane, alpha, through its center.

203

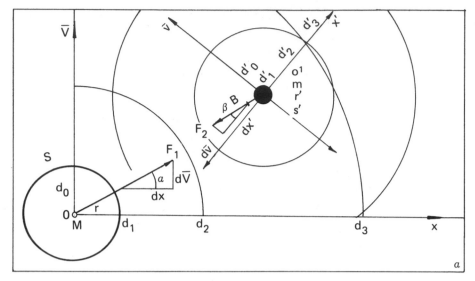

Figure 55

We add a second body that, having its own field, interferes with the first because of its moving in the neighborhood. Forces arise that continually modify equilibrium conditions (Figure 55).

The point O is influenced, for a time dt, by the gradient $d\overline{V}/dx$ (equal to tg α) and also, for the same dt, by the gradient $d\overline{v}/dx$ (equal to tg β).

F_1 and F_2 are colineal and in opposite directions, hence the values are subtracted and we have the approach value toward equipotential lines of higher value. This is not attraction; velocities are ceded.

If we assign a field value of 100 to O and at the surface S, we assign a value of 90, the value of the radius r of the body will be given by 100 – 90, or 10, at that distance.

If O' should coincide with O, the mass would be M + m and the velocity overall, $\overline{V} + \overline{v}$. In other words, the body of lesser mass would have ceded all its velocity to that of greater mass, and together they would form a new unit, the total mass and velocity of which would be equal to the sum of those contributed.

We obtain the distances d_i from the expression

$$F_i = \pm\, G\, \frac{M.m.\, \overline{V}}{d^2\, \overline{v}}$$

$$d_i = \sqrt{G\, \frac{M\, m\, \overline{V}}{F_i\, \overline{v}}}\ .$$

204

If there is no second body, there is no F, only a definable potential field.

(b) THE CASE OF THREE BODIES

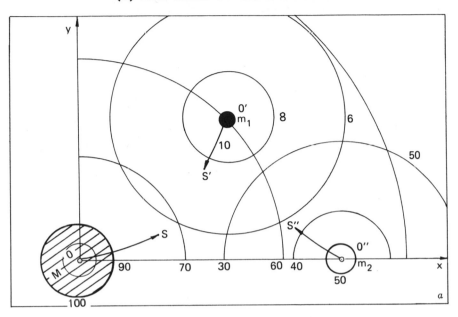

Figure 56

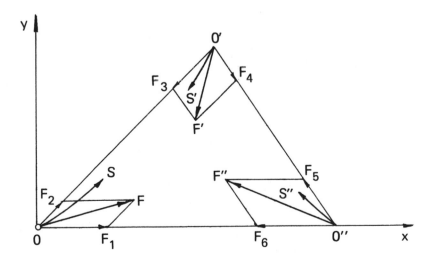

Figure 57

205

O moves with F   the trajectory of O is s
O moves with F′   the trajectory of O′ is S′
O moves with F″   the trajectory of O″ is S″

(c) THE CASE OF THREE BODIES IN SPACE

The coordinates of the centers with reference to the origin are

$$O \;\; (x, y, z)$$
$$O' \;\; (x', y', z')$$
$$O'' \;\; (x'', y'', z'').$$

The trajectories S_i are calculated with the F_i, M_i, m_i, $\overline{V}_i$, $\overline{v}_i$ and integrated over a line. All the bodies transfer velocities between one another. The final momentum is equal to the initial one.

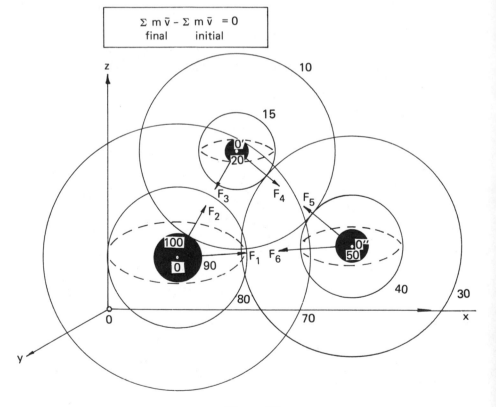

Figure 58

206

Appendix B

SUMMARY OF EXPRESSIONS IN THE UNIFIED FIELD

Usual

G'	$K.G$	$h.c^2,\ \dfrac{h}{\mu_0\,\epsilon_0},\ \sqrt{\dfrac{E.\lambda.v.h}{\mu_0.\epsilon_0}},\ \dfrac{E.h.v^2}{2q.V},\ G.K_p.R,\ \text{etc.}$
G	$\dfrac{F.d^2}{M.m}$	$\dfrac{h.c^2}{K_p.R},\ \dfrac{h}{K_p.\mu_0.\epsilon_0.R},\ \dfrac{K}{G'},\ \text{etc.}$
K	$\dfrac{G'}{G}$	$\dfrac{h}{G.\mu_0.\epsilon_0},\ \dfrac{h.c^2}{G},\ \dfrac{v^2.h^3}{m^2.G},\ K_p.R,\ \text{etc.}$
K_p	$\dfrac{K}{R}$	$\dfrac{G'}{G.R},\ \dfrac{h.c^2}{GR},\ \text{etc.}$
R	$\dfrac{K}{K_p}$	$\dfrac{G'}{G.K_p},\ \dfrac{G'.R}{G.K},\ \dfrac{h.R}{\mu_0.\epsilon_0.G.K},\ \text{etc.}$
F	$G\dfrac{M.m}{d^2}$	$\pm\,G\dfrac{M.m}{d^2}\dfrac{\overline{V}}{\overline{v}},\ \pm\left(\dfrac{h^2.E_1E_2.\overline{V}}{(K_p.R)^2G.d^2.\overline{v}}\right),\ \text{etc.}$
F	$\pm\,C\dfrac{Q.q}{d^2}$	$\pm\,C\dfrac{M.m}{d^2.\overline{v}}\overline{V},\ \pm\dfrac{C}{G}\left(\dfrac{h^2.E_1E_2\,\overline{V}}{(K_p.R)^2.G.d^2\overline{v}}\right),\ \text{etc.}$
g	$G\dfrac{M}{d^2}$	$\dfrac{h.c^2.M}{K_p.d^2.R},\ \text{etc.}$
p	$m.g$	$\dfrac{E.h^2.c^2M}{G.K^2.d^2},\ \dfrac{v.M}{c.K.d^2},\ \dfrac{v.M}{c.K_p.R.d^2},\ \text{etc.}$
m	$\dfrac{E}{c^2}$	$\dfrac{K_p.R.G.\mu_0.\epsilon_0}{\lambda.v},\ \dfrac{E.h}{G.K_p.R},\ \text{etc.}$
E	$m.c^2$	$\dfrac{M.G.K_p.R}{h},\ \dfrac{G.K_p.\mu_0.\epsilon_0.R}{\lambda.v.h},\ \text{etc.}$
c	$v\lambda,\ \sqrt{\dfrac{m}{E}},\ \sqrt{\dfrac{1}{\mu_0.\epsilon_0}}$	$\sqrt{\dfrac{GK_pR}{h}},\ \sqrt{\dfrac{g.R.K_p.d^2}{h.M}},\ \left(\dfrac{K_p.G.v.R}{m}\right)^{-\frac{1}{4}},\ \text{etc.}$

Usual

h	$\lambda.m.v$ $\dfrac{E}{\nu}$	$\dfrac{R.K_p.G}{c^2}$, $\dfrac{R.g.K_p.d^2}{c^2.M}$, $\sqrt{\dfrac{G.K_p.m.R}{\nu}}$, $\mu_0.\epsilon_0.K_p.R.G,$ $\mu_0.\epsilon_0.G'$, $\dfrac{R.G.K_p.\mu_0.\epsilon_0}{E.\lambda.v}$, etc.
λ	$\dfrac{h}{m.v}$ $\dfrac{c}{\nu}$	$\dfrac{G.K_p.R}{c^2.m.v}$, $\sqrt{\dfrac{R.K_p.G}{m.v^2.\nu}}$, $\dfrac{G^{3/2}.K^{3/2}}{h^{1/2}.c^4.m}$, etc.
ν	$\dfrac{E}{h}$	$\dfrac{R.G.K_p.m}{h^2}$, $\dfrac{c^4.m}{K_p.G.R}$, $\dfrac{K.c}{K_p}\left(\dfrac{1}{2^2}-\dfrac{1}{n^2}\right),$ $\dfrac{h.c}{G\mu_0.\epsilon_0 K_p}\left(\dfrac{1}{2^2}-\dfrac{1}{n^2}\right)$, $\dfrac{h.c^3}{G.K_p}\left(\dfrac{1}{2^2}-\dfrac{1}{n^2}\right),$ $\dfrac{\nu^2.h^2.c}{m^2.G.K_p}\left(\dfrac{1}{2^2}-\dfrac{1}{n^2}\right)$, etc.

Appendix C

EXERCISES

EXERCISE 1

When we look at the case of a photon wavetrain having many photons of different energies, we saw that on passing through a potential field (prism) or a large series of potential fields over long periods of time in space at 300,000 km/sec, a classification took place. The light finished up having a greater wavelength and a lower frequency—in other words, less energy.

If, as appears to be the case, it has a rotating element, the increase in wavelength would mean that the rotating element moves to an orbit more distant from the central nucleus. This, though, would mean a higher energy level, not lower, and the photon overall would be in a higher energy state. What, in fact, is going on?

In binaries, the orbits are reciprocal and this is a singular phenomenon. On a reduction of the energy of the rotating bodies, they move to orbits at a lower equipotential—i.e., more distant orbits. The phenomenon is that in binary systems, lower energy corresponds to greater separation. The reciprocal equilibrium state is set by equipotentials which are further removed than those corresponding to a higher energy.

In the diagram below, we have the normal condition of a two-body system where the respective masses are very different from one another.

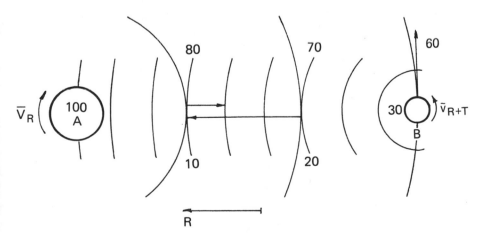

Figure 59

209

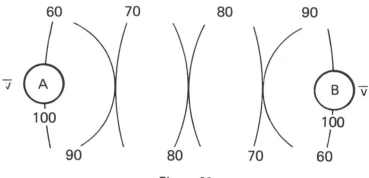

Figure 60

The $\bar{v}$ of B of greater energy moves to an A-based equipotential of lower energy but its $\bar{v}_{R\text{-}T}$ has to increase, thus increasing the energy.

If we now look at a diagram for a binary system, where masses and velocities are the same, then for an energy loss we would find an increased separation and greater wavelength with the corresponding lower frequencies and velocities (*i.e.*, energy).

As the energy, E, diminishes, so too do the values $\bar{v}_A$ and $\bar{v}_B$ (which are equal), the bodies passing to an equilibrium position at lower value reciprocal equipotentials; velocity is lower, the distance covered (in rotation) is greater, and frequency drops while wavelength increases. All this would support the photon's being a binary system.

The energy loss arises from the binray pair. If this, in fact, were the case, radio waves should have a speed less than that of light. Something else thrown up, in the same vein, is that radio waves would have a corpuscular basis. Could this be possible? Would those who have stuck it out as far as this point in the essay, having taken in all the points, deny outright that radio waves may have a corpuscular basis? That is to say, would be corpuscle have the characteristics of the organization of matter corresponding to its (lower) domain?

Our mental inertia always tends to have us imagine matter in any domain at the same as that with which we are familiar in U. With the necessity of discovering its properties in other domains, however, we will finish up having to define its structure in those domains.

EXERCISE 2

When, in Chapter XIII, we saw light as exerting a pressure and that a light-emitting body would suffer a thrust, it was to be supposed that under these circumstances, Newton's Third Law—all action is opposed by an equal and opposite reaction—would be complied with. In Chapter XIX, however, we considered the influence of the light source, with respect to the light emitted, irrelevant. Although this would be in conflict with Newton's Third Law and our second principle, we adopted as valid the Second Relativity Principle, whereby the velocity of light is independent of the velocity of its source.

Could the reader, in the light of the criteria set forth in this essay, analyze the situation with arguments for and against and justify the choice of principle made without thereby invalidating the first two principles mentioned?

EXERCISE 3

Can the following questions be answered in the affirmative?

(a) Are smaller, more active bodies comparatively more likely to give up velocity because they have more of it per unit mass?
(b) If this were valid for all domains, would it serve to show that energy comes from the lower domains?

EXERCISE 4

Since the equipotential surfaces of a spherical body alone and at rest in space are spherical, if it moves (revolves), these surfaces become elipsoids of revolution.

The first elipsoid would be the surface of the body itself. The impossibility, on revolving, of moving the entire mass uniformly, as in translational movement, comes about because the major axis of the elipsoid is more distant from the center of revolution than it is at the poles, the field being stronger in that direction.

In Figure 61 we see that at the same distance (above the pole) that the field has a value of 60, it has a value of 70 at the equator. Now, the field is stronger where there is more momentum—namely, in the equatorial plane.

The question is, why do bodies rotating about another do so in this plane?

That is to say that such bodies have a preference for moving in a path in AB (See Figure 62) rather than another, such as CD.

The explanation is that if we consider such a body as a test charge moving through space, to arrive at some time in the neighborhood of

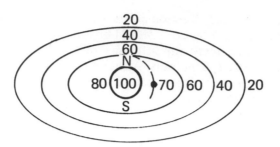

Figure 61

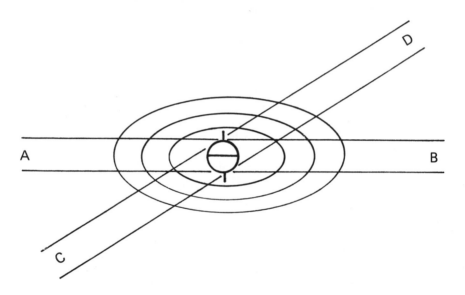

Figure 62

the main body, it will move toward the AB plane where the field is uniform, while to follow a path in CD would subject it to more or less sharp changes.

If the angle of incidence of the newly arrived body is such that it begins by following a CD-type trajectory (nonequatorial), then the interference gradients of its potential fields will guide it progressively into the AB plane.

Can the reader sketch a scheme of such resultants such that an incoming body shifts from any nonequatorial plane to the equatorial?

EXERCISE 5

The confusion problem: When we draw

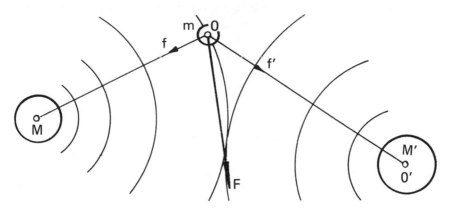

our mental tendency is to consider f and f′ as attraction forces and F
direction of movement; in fact they are: F is real force (m$\bar{v}$) and f and
f′ decomposition of F in "the interference of potential fields of M and
M′ (and m own that we consider negligable)."

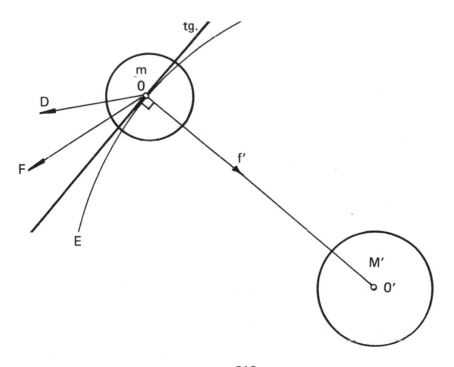

The answer why f' has the O–O' direction is by construction because the line of action of the component f' of F must be in a line normal to the tangent at point O of the equipotential E. All this is for instantaneous movement in a dt, from which the next direction could be D; then for the next dt we apply F to D direction and the process continues; then $\overline{v}$ becomes an acceleration and F = ma.

On that base, which is the situation with a mass m of F = O?

EXERCISE 6

Let's consider gravity waves. If a wave is produced by a perturbation of a medium, whence is a mechanical phenomena with certain parameters that define it and because there isn't in the spatial vacuum a perturbable medium but exist the potential fields interferences originating a mutual latent perturbation manifested with the presence of a test charge, on this base analyze the following:

(a) Is a dynamic process in a point of space if some body moves in the domain considered, the perturbation that reach that point is variable.

(b) Locating an infinitesimal test mass in such spatial point with F = O will be subject to oscillations because of the relative positions of all bodies in all space of such domain.

(c) If we consider a weighting scale to measure the same test mass on the earth surface, in what locations and in what times must we read the weights to detect differences to get values of the perturbation parameters?

(d) What characteristics must such perturbations have?

(e) In case that exist a positive answer, is possible consider such phenomena as of waves, by definition?

TABLE OF CONSTANTS

1 erg = 1 dyne x cm = gr cm^2/sec^2

1 Joule = 10^7 erg = 10^7 gr cm^2/sec^2

1 Kgm = 9.8 Joule = 9.8 x 10^7 erg

Mass of earth: 5.98 x 10^{27} gr

1 eV = 1.602 x 10^{-12} erg

1 MeV = 10^6 eV

Mass of proton = 1.672 x 10^{-24} gr

Mass of electron = 9.108 x 10^{-28} gr

1 Heat unit = 4.186 Joule

h = 6.6 x 10^{-26} gr cm^2/sec

C = 9 x 10^9 New m^2/Coul^2

G = 6.7 x 10^{-8} dyne cm^2/gr^2

K = 8.84 x 10^2 gr^2 cm/sec

c = 3 x 10^{10} cm/sec

G' = 5.6 x 10^{-5} gr cm^4/sec^3

ϵ_0 = 8.85 x 10^{-12} Coul^2/New m^2 = 8.85 x 10^{-5} gr sec^2/cm^3

μ_0 = 1.25 x 10^{-7} cm/gr

σ = 5.7 x 10^{-5}

k = 1.38 x 10^{-16} erg/molecule degree

K_p = 8 x 10^{-7} gr^2 cm^2/sec

R = 1.09 x 10^9 cm^{-1}

TABLE OF PARTICLES

Particle	Symbol	Mass (MeV)	Product of desintegration	Half life (seconds)
Photon	γ^0	?	stable	—
neutrin	ν^0	?	stable	—
electron	e^-	0,511	stable	—
muon	μ^-	105,6	$(e^-\nu\nu)$	$2,21\times10^{-5}$
pion	π^+	139,6	$(\mu^+\nu)$	$2,55\times10^{-8}$
	π^0	135,0	$(\gamma\gamma)$	1×10^{-16}
kaon	K^+	439,9	$(\mu^+\nu)(\Pi^+\Pi^0)(\mu^+\Pi^0\nu)$ $(e^+\Pi^0\nu)(\Pi^+\Pi^+\Pi^-)(\Pi^+\Pi^0\Pi^0)$	1.22×10^{-8}
	K_1^0	497,8	$(\Pi^+\Pi^-)(\Pi^0\Pi^0)$	1×10^{-10}
	K_2^0	497,8	$(\Pi^\pm e^\mp\nu)(\Pi^\pm\mu^\mp\nu)(\Pi^+\Pi^-\Pi^0)(3\Pi^0)$	$\sim6\times10^{-8}$
meson eta	η^0	548	$(\Pi^+\Pi^-\Pi^0)(3\Pi^0)(\Pi^+\Pi^-\gamma)(\gamma\gamma)$	short
meson rho	ρ^+	763	$(\Pi^+\Pi^0)$	short
	ρ^0		$(\Pi^+\Pi^-)$	short
meson omega	ω^0	782	$(\Pi^+\Pi^-\Pi^0)(\Pi^0\gamma)$	short
proton	p^+	938,2	stable	
neutron	n^0	939,5	$(p^+e^-\nu)$	$1,01\times10^3$
lambda	λ^0	1115,4	$(p\Pi-)(\eta\Pi 0)$	$2,5\times10^{-10}$
sigma	Σ^+	1189	$(\eta\Pi^+)(p\Pi 0)$	$0,8\times10^{-10}$
	Σ^0	1193	$(\lambda\gamma)$	short
	Σ^-	1197,4	$(\eta\Pi^-)$	$1,6\times10^{-10}$
cascade	Ξ^0	1315	$(\lambda^0\Pi^0)$	$2,8\times10^{-10}$
	Ξ^-	1321,2	$(\lambda^0\Pi^-)$	$1,75\times10^{-10}$

SYMBOLS

λ = wave length
c = speed of light
ν = frequency
h = Planck's constant
m = mass
v = velocity
E = energy
G = gravity constant of domain U
K = gravity constant of domain e
G' = total gravity constant
g = acceleration of gravity
d = radius of the earth
M = mass of the earth
μ = electrical permeability
ϵ = electrical induction
C = specific heat
Q = heat capacity
q = electric charge
V = voltage
F = force
p = pressure
ρ = density
v' = volume
L = length
$\bar{a}$ = characteristic acceleration
P = weight
T = temperature
k = Boltzman's constant
N = number of molecules
n = Avogadro number
A = area
R = Balmer's constant
K_p = ee domain own constant

Appendix D

TABLES OF WEIGHTLESSNESS

Rotating Body: Ring of radius r to center of rotation, centrifugal force is considered in the longitudinal direction of the ring body for strength of material.

In Table I we have: in column 1 the revolutions per minute and in column 2 the revolutions per second. In column 3 the distance that travel the center of mass for radius from 1 to 60 meters. In column 4 the speeds and in 5 the percentage of loss of weight related to the earth. In column 6 we have the remaining of loss of weight over 100% of the ring itself and we take this as capacity of the ring to pick up extra weight.

Table I

1	2	3	4	5	6
Rpm x 10³	Rps Rpm/60	dxturn	m/s (2x3)	% loss (4/300)	% wing to lift
		r: 1			
1	16.67	6.28	104.68	0.35	—
5	83.33	6.28	523.31	1.74	—
10	166.67	6.28	1046.68	3.49	—
15	250,-	6.28	1570,-	5.23	—
20	333.33	6.28	2093.31	6.97	—
		r: 5			
1	16.67	31.4	523.43	1.74	—
5	83.33	31.4	2616.56	8.72	—
10	166.67	31.4	5233.43	17.44	—
15	250,-	31.4	7850,-	26.16	—
20	333.33	41.4	10466.56	34.88	—
		r: 10			
1	16.67	62.8	1046.87	3.48	—
5	83.33	62.8	5233.12	17.44	—
10	166.67	62.8	10466.87	34.88	—
15	250,-	62.8	15700,-	52.33	—
20	333.33	62.8	20933.12	69.77	—

		r: 20			
1	16.67	125.6	2093.75	6.98	—
5	83.33	125.6	10466.25	34.88	—
10	166.67	125.6	20933.75	69.78	—
15	250,-	125.6	31400,-	104.67	4.67
20	333.33	125.6	41866.25	139.55	39.55
		r: 30			
1	16.67	188.4	3140.62	10.47	—
5	83.33	188.4	15699.37	52.33	—
10	166.67	188.4	31400.63	104.67	4.67
15	250,-	188.4	47100,-	157,-	57,-
20	333.33	188.4	62799.37	209.33	109.33
		r: 40			
1	16.67	251.2	4187.50	13.96	—
5	83.33	251.2	20932.49	69.74	—
10	166.67	251.2	41867.50	139.55	39.55
15	250,-	251.2	62800,-	209.33	109.33
20	333.33	251.2	83732.49	279.10	179.10
		r: 50			
1	16.67	314,-	5234.38	17.44	—
5	83.33	314,-	26165.62	87.21	—
10	166.67	314,-	52334.38	174.44	74.44
15	250,-	314,-	78500,-	261.66	161.66
20	333.33	314,-	104665.62	348.88	248.88
		r: 60			
1	16.67	376.80	6281.25	20.93	—
5	83.33	376.80	31398.74	104.66	4.66
10	166.67	376.80	62801.25	209.33	109.33
15	250,-	376.80	94200,-	314,-	214,-
20	333.33	376.80	125598.74	418.66	318.66

d : distance

In Table II we have: in column 1 the radius, in 2 the weights, in 3 how many Rpm must have the ring to loss his own weight, in column 4 we limit such quantity to 20,000 rpm and we calculate in column 5 how many kilograms brings to lift.

Table II

1	2	3	4	5	
r (m)	W (Kg) x10^3	Rpm (for 100% loss of w.) 1800000/2πr	Total (for 20 K Rpm)	Maximum of Kg. to lift	
5	2	57,325	less than W	0	
5	4	57,325	less than W	0	
5	6	57,325	less than W	0	
5	8	57,325	less than W	0	
5	10	57,325	less than W	0	
10	2	28,662	less than W	0	
10	4	28,662	less than W	0	
10	6	28,662	less than W	0	
10	8	28,662	less than W	0	
10	10	28,662	less than W		
15	2	19,108.28	less than W	0	
15	4	19,108.28	less than W	0	
15	6	19,108.28	less than W	0	
15	8	19,108.28	less than W	0	
15	10	19,108.28	less than W	0	
20	2	14,331.21	139.55	39.5%x2000:	791
20	4	14,331.21	139.55	39.5%x4000:	1,582
20	6	14,331.21	139.55		2,373
20	8	14,331.21	139.55		3,164
20	10	14,331.21	139.55		3,955
25	2	11,464.97	174.44	74.44%x2000:	1,488.80
25	4	11,464.97	174.44		2,977.60
25	6	11,464.97	174.44		4,466.40
25	8	11,464.97	174.44		5,955.20
25	10	11,464.97	174.44		7,444.–
30	2	9,554.14	209.33	109.33%x2K:	2,186.60
30	4	9,554.14	209.33		4,373.20
30	6	9,554.14	209.33		6,559.80
30	8	9,554.14	209.33		8,746.40
30	10	9,554.14	209.33		10,933.–

40	2	7,165.60	279.10	179.10%x2K:	3,582.-
40	4	7,165.60	279.10		7,164.-
40	6	7,165.60	279.10		10,746.-
40	8	7,165.60	279.10		14,328.-
40	10	7,165.60	279.10		17,910.-
50	2	5,732.48	348.88	248.88%x2K:	4,977.60
50	4	5,732.48	348.88		9,955.20
50	6	5,732.48	348.88		14,932.80
50	8	5,732.48	348.88		19,910.40
50	10	5,732.48	348.88		24,888.-

K:1000